DATE DUE

COMFORTABLE QUARTERS
for Laboratory Animals

Edited by
VIKTOR AND ANNIE REINHARDT

Ninth Edition, 2002
Animal Welfare Institute

To Christine Stevens

Her compassion is the inspiration behind this book.

Comfortable Quarters for Laboratory Animals

Ninth Edition, 2002

ISBN 0-938414-02-X
LCN 2002 141406

Animal Welfare Institute
PO Box 3650
Washington, DC 20007
http://www.awionline.org

This publication can also be accessed on the internet at:
http://www.awionline.org/pubs/cq02/cqindex.html

Cover illustration by Sämi Frei
Design by Ava Armendariz

Printed on recycled paper with soy ink.

Contents

i	Foreword *David B. Morton*
ii-iv	Introduction *Viktor and Annie Reinhardt*
1-5	The Ill-Effects of Uncomfortable Quarters *William M.S. Russell*
6-17	Comfortable Quarters for Mice in Research Institutions *Chris M. Sherwin*
18-25	Comfortable Quarters for Gerbils in Research Institutions *Eva Waiblinger*
26-32	Comfortable Quarters for Rats in Research Institutions *Monica M. Lawlor*
33-37	Comfortable Quarters for Hamsters in Research Institutions *Gernot Kuhnen*
38-42	Comfortable Quarters for Guinea-pigs in Research Institutions *Viktor Reinhardt*
43-49	Comfortable Quarters for Rabbits in Research Institutions *Boers K., Gray G., Love J., Mahmutovic Z., McCormick S., Turcotte N., and Zhang Y.*
50-55	Comfortable Quarters for Cats in Research Institutions *Irene Rochlitz*
56-64	Comfortable Quarters for Dogs in Research Institutions *Robert Hubrecht*
65-77	Comfortable Quarters for Primates in Research Institutions *Viktor Reinhardt*
78-82	Comfortable Quarters for Pigs in Research Institutions *Temple Grandin*
83-88	Comfortable Quarters for Sheep in Research Institutions *Viktor and Annie Reinhardt*
89-95	Comfortable Quarters for Cattle in Research Institutions *Viktor and Annie Reinhardt*
96-100	Comfortable Quarters for Horses in Research Institutions *Katherine A. Houpt and T.S. Ogilvie-Graham*
101-108	Comfortable Quarters for Chickens in Research Institutions *Detlef W. Fölsch, Marlene Höfner, Marion Staack and Gerriet Trei*
109-114	Comfortable Quarters for Amphibians and Reptiles in Research Institutions *Michael D. Kreger*

"The majority of scientists seem to make great efforts to avoid being associated with 'animal welfarists' or to become open to allegations of being somehow 'scientifically soft.' However, awareness of actual and potential stress and distress among animals in whatever situation should not be regarded as subjective but as a sound scientific base for the study of animals. Whether an observer maintains a high personal respect of the well-being of the individual animal or holds classic concepts of animals as being experimental 'models,' it should be more widely recognized that there is typically a scientific necessity to have animals at ease with their environments if studies are to remain objective."

—Clifford Warwick 1990. Important ethological and other considerations of the study and maintenance of reptiles in captivity. Applied Animal Behaviour Science 27, 363-366

Foreword

Professor David B. Morton
Head Centre for Biomedical Ethics, University of Birmingham, Edgbaston, Birmingham, B15 2TT, United Kingdom

I am truly delighted to be asked to write the foreword to this collection of essays. I have been using earlier editions as a guide as to how best I [as the animals' advocate and veterinarian] can protect the physiological and psychological integrity of the animals in my care and, therefore, their scientific utility. In this 9th edition, the editors have expanded the chapter on rodents. This is particularly timely given the tremendous increase in the production and use of transgenic mice, a trend that will continue for several decades. Also the chapter on chickens is substantially revised, and for those using other species of birds another recent review may be of interest (Hawkins et al., 2001).

Some years ago I defined 'Refinement' along the lines of minimising animal suffering during an experiment but in addition incorporating the notion of enhancing animal well-being through better husbandry. This is important as laboratory animals spend all of their lives in confinement of one sort or another. It should not be forgotten that their environment encompasses the effects that handlers and researchers have on the animals, which can be significant, and this has recently been reviewed (ILAR, 2001). There is now a plethora of evidence that confinement and husbandry conditions in general can, not surprisingly, impact on the physiological responses of animals in many sorts of experiments. Mind body separation is no more valid in animals than it is in humans, and the effects of psychological perturbations on physiological outcomes are considerable and ignored at the expense of producing poor science. Even when we are not sure of any impact on the animals, attempting to meet their needs and avoiding adverse mental or physical states, such as boredom, frustration, pain, distress, can only be to the benefit of both animals and science. This book addresses specifically what might count as good and poor environments for most of the common species used in laboratories. Each chapter not only provides, in the editors' words 'inspiration,' but also a wealth of practical ideas and guidance and those contributing to this book have provided us with a rich resource of information. By 'us' I mean animal care staff, veterinarians, scientists, inspectors and members of ethical review processes such as IACUCs.

The notion that it is of importance to animals that husbandry practices contribute to their well-being raises an ethical dimension to the debate. Poor husbandry and poor scientific practices of any sort will adversely affect their welfare, and if we can do it better then we are morally obliged to do so for ethical, legal, scientific and ultimately economic reasons. (NB, economic as poor welfare will increase the variance of data and so more animals will have to be used; worse still the results obtained may actually be invalid.) Consequently, it is useful to consider 'suffering,' such as pain and mental distress, that is 'avoidable' from suffering that is 'necessary' to achieve a scientific objective. The former is impermissible, the latter should be minimised. Consequently, if it is possible to avoid causing poor animal well-being, then we are obliged to do so and this book will help us do just that.

References

Hawkins P, Morton DB, Cameron D, Cuthill I, Francis R, Freire R, Gosler A, Healy S, Hudson A, Inglis I, Jones A, Kirkwood J, Lawton M, Monaghan P, Sherwin C, Townsend P 2001. Laboratory birds: refinements in husbandry and procedures: Fifth report of BVAAWF/FRAME/RSPCA/UFAW Joint Working Group on Refinement. Laboratory Animals 35(Supplement 1), 1-163

ILAR Journal 20007. Institute of Laboratory Animal Resources (National Research Council) Implications of Human-Animal Interactions and Bonds in the Laboratory. ILAR Journal 43(1) [Contributors: Thomas L. Wolfle, Kathryn Bayne, Fon T. Chang, Lynette A. Hart, Hank Davis, Harold Herzog, Lilly-Marlene Russow, Susan A. Iliff]

Introduction

Viktor and Annie Reinhardt
Animal Welfare Institute, PO Box 3650, Washington, DC 20007

Is it really necessary to provide animals in research institutions with comfortable quarters? Yes, species-adequate housing and handling conditions are not only a safeguard for the well-being of the animals but also a prerequisite for sound scientific methodology. Inadequacy of animal care can skew scientific findings and render the particular research useless (Donnelley and Nolan, 1990). It would, indeed, be naïve to rely on data collected from an animal:

- who experiences discomfort, frustration and/or distress resulting from spatial restriction [e.g., enclosure is too small to allow free, species-typical posturing and postural adjustment];
- who experiences discomfort, pain, fear, anxiety and/or distress resulting from enforced bodily restraint [e.g., immobilization during procedures]; or
- who experiences depression and frustration resulting from the inability to show species-typical behaviors [e.g., social animals kept in barren single-cages/stalls].

These experiences are reflected in an animal's physiological, psychological and behavioral responses to an experimental situation. The responses, however, differ from animal to animal because the experience itself is subjective. It is impossible to do truly "scientific" research under such methodological conditions because the data collected are influenced by unaccounted-for extraneous variables such as distress, fear, anxiety, discomfort, depression and boredom. "To demonstrate any experimental response against such a variable background generates a requirement for greater animal usage if the result is to be statistically valid" (Home Office, 1989, p. 8). "Good husbandry minimizes variations that can modify an animal's response to experimentation" (National Research Council, 1985, p. 11), thereby allowing the use of fewer animals giving equally valid results (Russell and Burch, 1959; Brockway et al., 1993; Chance and Russell, 1997). It is a fundamental scientific principle that all variables that have not proven to be insignificant be controlled in order to assure a sufficiently high degree of accuracy and reproducibility of the research findings. "If a researcher, through carelessness or ignorance, should use more animals for a project than is necessary, it must be considered unethical" (Öbrink and Rehbinder, 1999, p. 122).

Comfortable Quarters for Laboratory Animals offers suggestions and recommendations for how extraneous, husbandry-related variables can be minimized or avoided thereby maximizing the research animals' well-being and reducing the number of subjects required to obtain reliable research data. The basic conditions for the provision of comfortable quarters are outlined in regulations and professional guides:

Primary Enclosure

"Proper care, use, and humane treatment of animals used in research, testing, and education…require scientific and professional judgement based on knowledge of the needs of the animals….A good management program provides the environment, housing, and care that…minimizes variations that can affect research….The environment in which animals are maintained should be appropriate to the species….Animals should be housed with the goal of maximizing species-specific behaviors and minimizing stress-induced behaviors. For social species, this normally requires housing in compatible pairs or groups. …At a minimum, an animal must have enough space to turn around and to express normal postural adjustments …and must have enough clean-bedded or unobstructed area to move and rest in….Space allocations should be re-evaluated to provide for enrichment" (National Research Council, 1996, pp. 8, 22, 25 & 27).

Handling Procedures

"Handling of all animals shall be done as expeditiously and carefully as possible in a manner that does not cause …behavioral stress, physical harm, or unnecessary discomfort" (United States Department of Agriculture, 1995b, p. 21-22). "Restraint procedures should only be invoked after all other less stressful procedures have been rejected as alternatives….Physiological, biochemical and hormonal changes occur in any restraint animal…and investigators should consider how these effects will influence their proposed experiments" (Canadian Council on Animal Care et al., 1993, p. 95). "To reduce the stress and pain of laboratory animals, nontraumatic restraining techniques must be taught….We believe that teaching of procedural skills is crucial for maintaining high research standards within the laboratory" (Schwindaman, 1991, p. 30). "Many dogs, nonhuman primates…and other animals can be trained, through use of positive reinforcement, to present limbs or remain immobile for brief procedures" (National Research Council, 1996, p. 11).

Animal Care Personnel

"The behaviour of an animal during a procedure depends on the confidence it has in its handler. This confidence is developed through regular human contact and, once established, should be preserved….All staff, both scientific and technical, should be sympathetic, gentle and firm when dealing with the animals" (Home Office, 1989, p. 16-17).

Providing animals in research institutions comfortable, i.e., humane quarters is not only a scientific but also an ethical obligation. After all, the caged animal is completely at the mercy of the investigator. To merely "use" animals for personal gain [e.g., promoting one's academic career] or for perceived benefits for people [e.g., developing treatments of diseases] without paying proper attention for their safety and well-being is ethically not acceptable. To show concern for the well-being of research animals, however, may stigmatise an investigator as being "scientifically soft" even though "awareness of actual and potential stress and distress among animals in whatever situation should not be regarded as subjective but as a sound scientific base for the study of animals. Whether an observer maintains a high personal respect for the well-being of the individual animal or holds classic concepts of animals as being experimental 'models,' it should be more widely recognized that there is typically a scientific necessity to have animals at ease with their environments if studies are to remain objective" (Warwick, 1990, p. 363).

The chapters of the new edition have been written by animal care personnel, scientists and veterinarians who have demonstrated an active commitment to the humane and scientifically acceptable housing and handling of laboratory animals. In our invitation letter we have asked each author to:

- outline the species-typical characteristics of the species you are dealing with,
- make recommendations on how these characteristics can best be addressed in the research institution,
- make the well-being of the animals an uncompromising priority of your chapter and
- provide supportive references for all statements.

In the United States more than 14 million animals are used annually in research institutions. Only approximately 10% of these animals are regulated under the *Animal Welfare Act* (United States Department of Agriculture, 2000). The remaining 90% are either not considered at all [cold-blooded animals] or explicitly excluded [rats, mice, birds] in the regulatory definition of the term "animal" (United States Department of Agriculture, 1995a, p. 1), and they are, therefore, exempt from legal protection to "insure that animals intended for use in research facilities…are provided humane care and treatment" (Animal Welfare Act, 1985, p. 1). We see no scientific, ethical or logical justification for this seemingly arbitrary discrimination. Since rats and mice far outnumber all so-called "true animals" taken together, their inhumane care and treatment causes much more suffering and affects scientific findings in a much more pervasive manner. We feel that rats and mice, but also birds and cold-blooded animals, such as reptiles and amphibians, deserve the same consideration as other animals legally do, and we have therefore included chapters specifically addressing their needs for well-being in the research institution.

This is the ninth edition of *Comfortable Quarters for Laboratory Animals*, which was first published in 1955 for free distribution by the Animal Welfare Institute. May the recommendations set forth in this book serve as an inspiration to all those who are committed to safeguarding the well-being of research animals and the integrity of sound scientific methodology.

References

Animal Welfare Act as Amended (7 USC 2131-2156) 1985. **Full Text:** http://www.nal.usda.gov/awic/legislat/awa.htm

Brockway BP, Hassler CR, Hicks N 1993. Minimizing stress during physiological monitoring. In Refinement and Reduction in Animal Testing Niemi SM, Willson JE (eds), 56-69. Scientists Center for Animal Welfare, Bethesda, MD

Canadian Council on Animal Care, Olfert ED, Cross BM, McWilliam AA 1993. Guide to the Care and Use of Experimental Animals, Volume 1, 2nd Edition. Canadian Council on Animal Care, Ottawa, Canada
Full Text: http://www.ccac.ca/guides/english/toc_v1.htm

Chance MRA, Russell WMS 1997. The benefits of giving experimental animals the best possible environment. In Comfortable Quarters for Laboratory Animals, Eighth Edition Reinhardt V (ed), 12-14. Animal Welfare Institute, Washington, DC

Donnelley S 1990. Animals in science: The justification issue. In Animal, Science and Ethics Donnelley S, Nolan K (eds), 8-13. The Hastings Center, Garrison, NY

Home Office 1989. Animals (Scientific Procedures) Act 1986. Code of Practice for the Housing and Care of Animals Used in Scientific Procedures. Her Majesty's Stationery Office, London, UK
Full Text: http://www.homeoffice.gov.uk/animact/hcasp.htm

National Research Council 1985. Guide for the Care and Use of Laboratory Animals, 6th Edition. National Institutes of Health, Bethesda, MD

National Research Council 1996. Guide for the Care and Use of Laboratory Animals, 7th Edition. National Academy Press, Washington, DC
Full Text: http://www.nap.edu/readingroom/books/labrats

Öbrink KJ, Rehbinder C 1999. Animal definition: a necessity for the validity of animal experiments? Laboratory Animals 22, 121-130

Russell WMS, Burch RL 1959. The Principles of Humane Experimental Technique. Methuen & Co., London, UK
Full Text: http://altweb.jhsph.edu/publications/humane_exp/het-toc.htm

Schwindaman D 1991. The 1985 animal welfare act amendments. In Through the Looking Glass. Issues of Psychological Well-being in Captive Nonhuman Primates Novak MA, Petto AJ (eds), 26-32. American Psychological Association, Washington, DC

United States Department of Agriculture 1995a. 9 Code of Federal Regulations Ch. 1 (1-1-95 Edition) U.S. Government Printing Office, Washington, DC
Full Text: http://www.access.gpo.gov/nara/cfr/waisidx_00/9cfr1_00.html

United States Department of Agriculture 1995b. 9 Code of Federal Regulations Ch. 1 (1-1-95 Edition) U.S. Government Printing Office, Washington, DC
Full Text: http://www.access.gpo.gov/nara/cfr/waisidx_00/9cfr2_00.html

United States Department of Agriculture 2000. <u>Animal Welfare Report–Fiscal Year 2000</u> U.S. Department of Agriculture–Animal Care, Riverdale, MD
Full Text: http://www.aphis.usda.gov/ac/awrep2000.pdf

Warwick C 1990. Important ethological and other considerations of the study and maintenance of reptiles in captivity. <u>Applied Animal Behaviour Science</u> 27, 363-366

The Ill-Effects of Uncomfortable Quarters

William M.S. Russell
Department of Sociology, University of Reading, Whiteknights, PO Box 218, Reading, Berks, RG6 2AA, United Kingdom

The Three "R"s of humane experimental technique are: **Replacement** of animals by insentient material (e.g., tissue cultures), **Reduction** in the number of animals used to obtain given information (e.g., by proper statistical design and analysis of experiments), and **Refinement** of procedures actually used to minimise the distress imposed on the animals.

Distress may be inflicted directly, as an unavoidable consequence of the experimental procedure employed, or contingently, as an inadvertent by-product of the use of the procedure, which is not necessary for its success, and is always detrimental to the object of the experiment. The incidence of contingent distress "will include the results of every conceivable kind of imperfection in the husbandry of laboratory animals," of which their housing is a major component. "Where chronic experiments over days or months are concerned, we cannot even in principle separate husbandry from the conduct of the experiment itself. For husbandry means keeping the animals alive and healthy for long periods, and this is an essential part of, say, a nutritional experiment" (Russell and Burch, 1959). "Maintenance methods must always be considered as a critical factor in animal experiments, and they must be documented in detail in the experimental design" (Claassen, 1994). It follows that the third "R"—Refinement—is concerned not only with minimising distress during experiments [e.g., by the use of analgesics], but with maximising comfort and well-being of the animals in husbandry.

The reasons for this are not only humanitarian. In all contexts, there is positive correlation between humaneness and scientific efficiency—good science is humane science (Russell and Burch, 1959)—but this is particularly obvious in the present context. The point has been made again and again: "Stressed animals do not make good research subjects" (American Medical Association, 1992; cited by Reinhardt and Reinhardt, 2000). "Good animal experimentation requires consideration not only of temperature and cleanliness, but also of social environment and environmental change. If these direct influences on experiments remain unrecognised or uncontrolled, the validity of research on such animals is to be questioned" (Anonymous, 1974). Or we may go back to the supreme pioneer of our subject, Charles Hume (cited by Poole, 1999): "It fortunately happens that the animals most suitable for scientific research are those that are healthy, tame, comfortable and contented."

Mental or behavioural content is as important as bodily comfort; in fact the two are inseparable. "The major discovery of anatomy and physiology in the last half-century has been that of the extraordinarily subtle, comprehensive and intimate linkages and interactions between the somatic nervous system, the organ of behaviour, the autonomic nervous system and the endocrine system, which control events within the body" (Russell and Burch, 1959). It was already clear in the 1950s which parts of the brain were chiefly involved in these linkages—the hypothalamus in all vertebrates and the limbic system in mammals. These connections are capable of "converting distress caused by the physical or social environment into physiological stress bound to disturb any experimental results....More is known now about the pathways to and from the limbic system, and the corticotropin-releasing factor in the hypothalamus, discovered in 1955, was isolated in 1981 and has since been the subject of numerous studies—some *in vitro*—and related substances have been found in lower vertebrates. All such findings have served only to amplify and fully confirm the very close linkages between the three systems already established in the 1950s" (Russell, 1997). In man this is the basis for the discipline of psychosomatic medicine, equally important in the veterinary context. Disturbances in behaviour, even if they appear to us quite slight, cause mental distress and hence physiological stress.

With all this in mind, we can now consider some of the ill-effects of various defects in the physical, behavioural and social environments of laboratory animals. These ill-effects from unsatisfactory husbandry and housing will evidently be equally harmful to animal welfare and biomedical science.

The Physical Environment

We can begin with light. "Rats, mice and rabbits are nocturnal or crepuscular" (Rose, 1994). Strong light may be harmful for them. When cages are piled in racks, there may be considerable differences among them in light intensity (Bellhorn,

1980). The ones at the bottom can find shaded areas, the ones at the top can only avoid excessive illumination by huddling under the food hopper (Rose, 1994). Strong light can increase adrenal weight in male rats and cause an increase in pre-weaning mortality in mice (Fox, 1986). Even a moderate light intensity can damage the rat retina and blind albino rats, who lack a protective retinal pigment (Weihe, 1976; Fox, 1986; Rose, 1994; Clough, 1999). Conversely, herbivorous lizards who lack direct access to sunlight may suffer Vitamin D_3 deficiency unless ultraviolet light is artificially provided (Kreger, 1997). Rather surprisingly, some fish in outdoor ponds are liable to be sun-burned, unless protected from excess sunlight (Reddacliff, 1994). This raises the obvious but important point that different species have different needs, and that a knowledge of the life and behaviour of a particular species in the wild is invaluable when designing its laboratory environment.

"Loud sounds, audible to humans, raise triglyceride levels and blood pressure while lowering glucose levels in rats. In rats, mice and hamsters, they cause convulsions" (Russell, 1997). They can damage the cochlea of guinea-pigs, and cause atherosclerosis in rabbits (Fox, 1986). "Sudden noises, such as the clatter of metal cages being cleaned, can cause a 100-200 percent increase in plasma corticosterone in rats" (Fox, 1986).

Some species can hear ultrasounds, and "ultrasounds may be a common feature of the laboratory and animal house environment. Of 39 sound sources monitored, 24 were found to generate ultrasound" (Sales et al., 1989). Examples are cage washers, vacuum hoses, high pressure hoses, ringing telephones, running taps, oscilloscopes, and video and computer monitors, which "gave particular cause for concern as they were entirely ultrasonic and therefore appeared to be 'silent'"(Sales et al., 1989). "Ultrasounds increase sodium excretion and may damage the cochlea in rodents" (Russell, 1997). A very important aspect of the ill-effects of loud sounds is that "resulting effects can include atypical responses to drugs many weeks or months after the sound exposure occurs" (Clough, 1999).

While light and sound will suffice as examples, obviously other physical variables are important, such as temperature and relative humidity. For instance "large [10-fold] variations in toxicity can occur with only small [4°C] changes in ambient temperature during the course of an experiment" on the cardiotoxicity of isoprenaline in rats. Low relative humidity promotes ringtail disease in rats (Clough, 1999). Obviously in the husbandry of aquatic species correct properties of the water are vital (temperature, pH, dissolved oxygen, etc.; Gilman, 1984).

The Behavioural Environment

In 1962, the Swiss zoologist Ernst Inhelder studied a number of different animal species kept in impoverished environments, that is, simple cages or enclosures with none of the varied features and objects encountered by these species in the wild (Inhelder, 1962). It is worth pointing out that simulations of such features can be perfectly satisfactory. Heini Hediger, the distinguished ethologist and director of the Zürich Zoo, when in Africa, noticed that termite nests had their tops polished or rubbed away, because zebras came from miles away to rub against them as an important means of grooming their coats. On return he set up a cement model of a termite nest in the zebra enclosure of his zoo. The animals were so excited that they rushed to the "nest" and rubbed against it with such enthusiasm they knocked it over! Hediger made a reinforced model, which, he reported, "has been in daily use ever since" (Hediger, 1955).

The animals studied by Inhelder, less fortunate in their totally impoverished environments, had no opportunity to carry out many of their species-typical behaviour patterns, for lack of environmental features or objects to use. In these conditions they carried out instead repetitive stereotyped meaningless activities, such as walking back and forth a short distance so precisely as literally to tread in their own footsteps. This stereotypy resembles the human mental illness of compulsion neurosis, in which the sufferer feels compelled to repeat over and over again some act such as turning a gas jet off and on, or opening and closing a window (Fenichel, 1945). All accounts of this condition agree in showing that it is extremely distressing for the sufferer (Fenichel, 1945; Pfister, 1948; Freud, 1950). We may assume that stereotyped activity is equally distressing for the animals.

Inhelder's studies were made in zoos, but similar observations have been made on laboratory animals. Stereotyped activities in impoverished laboratory environments have been observed in rabbits (head swaying, biting bars, circling walk, pushing hopper with head, etc.; Morton et al., 1993), carnivores ("rocking, pacing, weaving and whirling"; Fox, 1986), rodents (Baenninger, 1967; Wiedenmayer, 1987; Würbel et al., 1998; Callard et al., 2000; Reinhardt and Reinhardt, 2001a), primates (Erwin and Deni, 1979; Poole, 1988; Harris, 1989) and birds (Morris, 1966). Increasing the cage size does not reduce stereotypy (Bayne and McCully, 1989; Galef and Durlach, 1993), but there is actually no reason why it should—an unfurnished large cage is just as impoverished an environment as an unfurnished small cage (Reinhardt and Reinhardt, 2001a). If cages are enriched with furnishings, large ones are better because they enhance the animals' behavioural health (Brent, 1992; Kitchen and Martin, 1996; Williams et al., 2000). In macaques and baboons, stereotypy is reduced or stopped by supplying "one or more perches… and other cage furniture," especially "a woodchip substrate which includes food items…as provisioned substrate for foraging" (Harris, 1989). The benefits of such enrichment show that the stereotypy is due to lack of opportunities for carrying out normal behaviour patterns, such as climbing, perching, foraging etc.

Since stereotypy is distressing, we must expect physiological repercussions of impoverishment. Sure enough, gerbils in an impoverished environment have higher cortisol levels than those in an enriched one (van de Weerd et al., 1996). An interesting comparison has been made between mice in a totally impoverished environment and mice supplied with materials they can use to make nests (van de Weerd et al., 1996). The mice in impoverished conditions ate more but weighed <u>less</u> than those supplied with nest materials, probably because they could not regulate their body temperature—by sleeping in a nest—without consuming more food.

We have seen that in rat-cage racks the rats at the top may get too much light. In double-tier macaque housing the monkeys in the lower cages may be in such dim light that "animal care staff routinely use flashlights to identify animals" (Reinhardt and Reinhardt, 2001b). Except for owl monkeys, apes and monkeys in general are active in daylight. A dark cage is as impoverished an environment for them as a dark dungeon for a human being. Not surprisingly, they show stereotyped acts in this situation.

The Social Environment

In social animals, including many mammal species, the social environment may be unsatisfactory in two different ways: crowding [too many animals in a given area] or isolation [animals caged singly]. Claire Russell and I described the many kinds of behaviour disturbance that occur in crowded and isolated monkeys (Russell and Russell, 1985). There is excessive frequency and intensity of several kinds of conflict activities, acts that in relaxed spacious conditions resolve momentary conflicts, but under crowding or isolation become repetitive and stressful. Under crowding these acts are socially disturbing, for instance rough grooming, or kidnapping babies. Under isolation, they tend to be self-directed—for obvious reasons—for instance repetitive stereotyped self-grooming in adults, or violent self-rocking in young monkeys [instead of the smooth motion of being carried by an adult].

However, the most prominent consequence of crowding or isolation for behaviour is violence. Under crowding conditions this means social violence, including the killing of females and young in particular. We showed that this change in behaviour, and the resulting physiological effects to be considered presently, are aspects of an overpopulation crisis response—a means of reducing populations before they deplete their resources (Russell and Russell, 1968, 1984, 1999). But of course this does not mean it is agreeable for the monkeys. Isolated monkeys redirect violence against themselves. They "pinch the same patch of their own skin repeatedly until it is raw, or even bite and tear themselves" (Russell and Russell, 1985). "Approximately 10% of captive, individually-housed monkeys engage in the disturbing phenomenon of self-injurious behavior" (Jorgensen et al., 1996) with self-inflicted lacerations needing veterinary attention. It has been estimated that there are approximately 15,000 individually-caged macaques in the United States alone, so that as many as 1,500 may be seriously injuring themselves (Reinhardt and Rossell, 2001).

Besides wounds inflicted by others [under crowding] or by themselves [under isolation], crowded and isolated animals are physiologically impaired. By its action on the adrenal glands, crowding causes "not only stress diseases but increased susceptibility to poisons, radiation, parasites and infections" (Russell, 1997). Evidence of the physiological effects of crowding "is available for mice, rats, woodchucks, rabbits, dogs and sika deer" (Russell and Russell, 1968).

Isolation also affects the adrenal glands, and it can cause excessive eating and drinking in rhesus monkeys, decreased leukocyte counts in mice, decrease in circulating lymphocytes and alterations in response to drugs and poisons in rats (Fox, 1986). Isolation can also cause increased blood pressure and hypertension in rats (Claassen, 1994). In macaques, it can increase the susceptibility to diarrhea and promote the development of coronary atherosclerosis (Shively et al., 1989; Schapiro et al., 2000). "Caging monkeys in isolation causes a decline in the number and function of the T cells so vital for immunity, and they recover when the monkeys are restored to their companions" (Russell, 1999). It will be obvious by now that many if not most singly-caged social animals, including the 15,000 macaques, are virtually useless for scientifically sound experimental purposes.

There is one final interesting point about the social environment. In 1956, Michael Chance studied the test response of immature female rats to serum gonadotrophin and found that "the variation in ovary weight [the test response] was greater if the rats were caged either singly or in groups larger than two [with the floor area per rat roughly constant]" (Chance and Russell, 1997). In 1992, Chance's colleague John Mackintosh obtained a similar result in a study of the response of male mice to a barbiturate anaesthetic. It follows that an optimal social environment can make animals more uniform, and the less animals vary in their responses the fewer of them we need to get a statistically representative sample. Here, therefore, reduction goes hand in hand with refinement.

Home or Away

There are a number of routine procedures that form part of many experiments: physiological monitoring, drug administration and blood collection. These can be carried out without restraining the animals or moving them from their home cage, by telemetry (Brockway et al., 1993; Schnell and Wood, 1993; Kramer, 2000) or training, for instance teach primates actively to cooperate during injection and venipuncture in their homecage (Priest, 1991; Reinhardt and Cowley, 1992). Unfortunately, instead of these refinements being used, animals are often transferred from their cages to some other place, where they are put under enforced restraint for these procedures. "Transfers between cages or containers raise corticosterone levels and cause weight loss in mice, and raise catecholamine and other hormone levels and susceptibility to the poison ammonium diuranate in rats. Simple tethering raises the heart rate in cynomolgus monkeys. More severe physical restraint raises the blood pressure in rats and mice, catecholamine levels in rats, and rectal temperature and pulse rate in golden marmosets. In rats, it can also affect numbers of receptors at certain sites and therefore directly affect responses to drugs" (Russell, 1997). In several species enforced immobilization can produce gastric ulceration (Fox, 1986).

Conclusion

From all these findings, it will now be obvious that the provision of comfortable quarters—including handling procedures—with a stable environment, companionship and freedom to engage in the species-typical basic activities, is of supreme importance, alike for laboratory animal welfare and for the progress of reliable biomedical science.

References

American Medical Association 1992. Use of Animals in Biomedical Research—The Challenge and Response—An American Medical Association White Paper. AMA. Group on Science and Technology, Chicago, IL (Cited by Reinhardt and Reinhardt, 2000)

Anonymous 1974. Review of Environmental Variables in Animal Experimentation (edited by Magalhaes, H), Bucknell University Press, Lewisburg, PA. ILAR [Institute of Laboratory Animal Resources] News 18(1), 20

Baenninger LP 1967. Comparison of behavioural development in socially isolated and grouped rats. Animal Behaviour 15, 312-323

Bellhorn RW 1980. Lighting in the animal environment. Laboratory Animal Science 30, 440-450

Brent L 1992. The effects of cage size and pair housing on the behavior of captive chimpanzees. American Journal of Primatology 27, 20

Brockway BP, Hassler CR, Hicks N 1993. Minimizing stress during physiological monitoring. In Refinement and Reduction in Animal Testing Niemi SM, Willson JE (eds), 56-69. Scientists Center for Animal Welfare, Bethesda, MD

Callard MD, Bursten SN, Price EO 2000. Repetitive backflipping behaviour in captive roof rats (Rattus rattus) and the effect of cage enrichment. Animal Welfare 9, 139-152

Chance MRA, Russell WMS 1997. The benefits of giving experimental animals the best possible environment. In Comfortable Quarters for Laboratory Animals, Eighth Edition Reinhardt V (ed), 12-14. Animal Welfare Institute, Washington, DC

Claassen V 1994. Neglected Factors in Pharmacology and Neuroscience Research. Elsevier, Amsterdam, Netherlands

Clough G 1999. The animal house: design, equipment and environmental control. In UFAW Handbook on the Care and Management of Laboratory Animals, Seventh Edition, Volume 1 Poole T, English P (eds), 97-134. Blackwell Science, Oxford, UK

Erwin J, Deni R 1979. Strangers in a strange land: Abnormal behavior or abnormal environments? In Captivity and Behavior Erwin J, Maple T, Mitchell G (eds), 1-28. Van Nostrand Reinhold, New York, NY

Fenichel O 1945. The Psychoanalytical Theory of Neurosis WW Norton, New York, NY

Fox MW 1986. Laboratory Animal Husbandry: Ethology, Welfare and Experimental Variables. State University of New York Press, Albany, NY

Freud S 1950. Notes upon a case of obsessional neurosis (1909). In Collected Papers, Volume 3, 291-383. Hogarth Press and Institute of Psychoanalysis, London, UK

Galef Jr. BG, Durlach P 1993. Should large rats be housed in large cages? An empirical issue. Canadian Psychology 34, 203-207

Gilman J (ed.) 1984. Guide to the Care and Use of Experimental Animals, Volume 2 Canadian Council on Animal Care, Ottawa, Canada
Full Text: http://www.ccac.ca/guides/english/toc_v2.htm

Harris DHR 1989. Environmental enrichment and its effects on the individual. In Laboratory Animal Welfare Research—Primates Universities Federation for Animal Welfare [UFAW] (ed), 15-16. UFAW, Potters Bar, UK

Hediger H 1955. Studies of the Psychology and Behaviour of Captive Animals in Zoos and Circuses. Butterworths Scientific Publications, London, UK

Inhelder E 1962. Skizzen zu einer Verhaltenspathologie reaktiver Störungen bei Tieren. Schweizer Archiv für Neurologie, Neurochirurgie und Psychiatrie 89, 276-326

Jorgensen MJ, Novak MA, Kinsey J, Tiefenbacher S, Meyer JS 1996. Correlates of self-injurious behavior in monkeys. XVIth Congress of the International Primatological Society/XIXth Conference of the American Society of Primatologists, Abstract No. 767

Jorgensen MJ, Kinsey JH, Novak MA 1998. Risk factors for self-injurious behavior in captive rhesus monkeys (Macaca mulatta). American Journal of Primatology 45, 187

Kramer K 2000. Applications and Evaluation of Radio-Telemetry in Small Laboratory Animals. Doctoral Thesis, University of Utrecht, Utrecht, Netherlands

Kreger MD 1997. Laboratory housing for reptiles and amphibians. In Comfortable Quarters for Laboratory Animals, Eighth Edition Reinhardt V (ed), 32-40. Animal Welfare Institute, Washington, DC
Full Text: http://www.awionline.org/pubs/cq/three.pdf

Kitchen AM, Martin AA 1996. The effects of cage size and complexity on the behaviour of captive common marmosets, Callithrix jacchus jacchus. Laboratory Animals 30, 317-326
Full Text: http://www.lal.org.uk/pdf.htm

Morris D 1966. Abnormal rituals in stress situations: the rigidifaction of behaviour. Philosophical Transactions of the Royal Society Series B 251, 327-330

Morton DB, Jennings M, Batchelor GR, Bell D, Birke L, Davies K, Eveleigh JR, Gunn D, Heath M, Howard B, Koder P, Phillips J, Poole T, Sainsbury AW, Sales GD, Smith DJA, Stauffacher M, Turner RJ 1993. Refinements in rabbit husbandry. Second report of the BVAAWF/FRAME/RSPCA/UFAW joint working group on refinement. Laboratory Animals 27, 301-329
Full Text: http://www.lal.org.uk/pdffiles/rabbit.pdf

Pfister O 1948. Christianity and Fear. Allen & Unwin, London, UK (Transl. Johnston WH)

Poole TB 1988. Normal and abnormal behaviour in captive primates. Primate Report 22, 3-12

Poole T 1999. Introduction. In UFAW Handbook on the Care and Management of Laboratory Animals, Seventh Edition, Volume 1 Poole T, English P (eds), 1-3. Blackwell Science, Oxford, UK

Priest GM 1991. Training a diabetic drill (Mandrillus leucophaeus) to accept insulin injections and venipuncture. Laboratory Primate Newsletter 30(1), 1-4
Full Text: http://www.brown.edu/Research/Primate/lpn30-1.html#loon

Reddacliff GL 1994. Examples of species-specific considerations in the design of an environment—fish. In Improving the Well-being of Animals in the Research Environment Baker RM, Jenkin G, Mellor DJ (eds), 135-138. ANZCCART [Australian and New Zealand Council for the Care of Animals in Research and Teaching], Glen Osmond, SA, Australia

Reinhardt V, Cowley D 1992. In-homecage blood collection from conscious stumptailed macaques. Animal Welfare 1, 249-255
FT: http://www.awionline.org/Lab_animals/biblio/aw1blood.htm

Reinhardt V, Reinhardt A 2000. Blood collection procedure of laboratory primates: A neglected variable in biomedical research. Journal of Applied Animal Welfare Science 3, 321-333
FT: http://www.awionline.org/Lab_animals/biblio/jaaws2.html

Reinhardt V, Reinhardt A 2001a. Legal space requirement stipulations for animals in the laboratory: Are they adequate? Journal of Applied Animal Welfare Science 4, 143-149
FT: http://www.awionline.org/Lab_animals/biblio/jaaws3.html

Reinhardt V, Reinhardt A 2001b. Environmental Enrichment for Caged Rhesus Macaques (Macaca mulatta)—Photographic Documentation and Literature Review (Second Edition). Animal Welfare Institute, Washington, DC
FT: http://www.awionline.org/lab_animals/rhesus/Photo.htm

Reinhardt V, Rossell M 2001. Self-biting in caged macaques: Cause, effect and treatment. Journal of Applied Animal Welfare Science 4, 285-294
FT: http://www.awionline.org/Lab_animals/biblio/jaaws4.html

Rose MA 1994. Environmental factors likely to impact on animal's well-being—an overview. In Improving the Well-being of Animals in the Research Environment Baker RM, Jenkin G, Mellor DJ (eds), 99-116. ANZCCART [Australian and New Zealand Council for the Care of Animals in Research and Teaching], Glen Osmond, SA, Australia

Russell WMS, Burch RL 1959. The Principles of Humane Experimental Technique. Methuen, London, UK
Full Text: http://altweb.jhsph.edu/publications/humane_exp/het-toc.htm

Russell C, Russell WMS 1968. Violence, Monkeys and Man. Macmillan, London, UK

Russell C, Russell WMS 1984. Overpopulation crisis. Social Biology and Human Affairs 49, 23-42

Russell C, Russell WMS 1985. Conflict activities in monkeys. Social Biology and Human Affairs 50, 26-48

Russell WMS 1997. Shooting the clock: Timeless lessons of the past still guide today's refinement initiatives. Science and Animal Care [WARDS Newsletter] 8(3), 1-2 & insert

Russell, WMS 1999. Reduction and refinement in biomedical experiments. Research Defence Society [RDS] News January, 12-13

Russell C, Russell WMS 1999. Population Crises and Population Cycles. The Galton Institute, London, UK

Sales GD 1989. Effects of environmental ultrasound on behaviour of laboratory rats. In Laboratory Animal Welfare Research—Rodents Universities Federation for Animal Welfare [UFAW] (ed), 7-16. UFAW, Potters Bar, UK

Schapiro SJ, Nehete PN, Perlman JE, Sastry KJ 2000. A comparison of cell-mediated immune responses in rhesus macaques housed singly, in pairs, or in groups. Applied Animal Behaviour Science 68, 67-84

Schnell CR, Wood JM 1993. Measurement of blood pressure, heart rate, body temperature, ECG and activity by telemetry in conscious unrestrained marmosets. Proceedings of the Fifth Federation of European Laboratory Animal Science Associations (FELASA) Symposium, 107-111

Shively CA, Clarkson TB, Kaplan JR 1989. Social deprivation and coronary artery atherosclerosis in female cynomolgus monkeys. Atherosclerosis 77, 69-76

van de Weerd HA, van Loo PLP, van Zutphen LFM, Koolhaas JM, Baumans V 1996. Long-term behavioural and physiological effects of nesting material as environmental enrichment for laboratory mice. In Environmental Enrichment for Laboratory Mice: Preferences and Consequences van de Weerd, 87-104. Doctoral Thesis, University of Utrecht, Netherlands

Weihe WH 1976. The effect of light on animals. Laboratory Animal Handbooks 7, 63-76

Wiedenmayer C 1997. The early ontogeny of bar-gnawing in laboratory gerbils. Animal Welfare 6, 273-277

Williams LE, Steadman A, Kyser B 2000. Increased cage size affects Aotus time budgets and partner distances. American Journal of Primatology 51, Supplement 1, 98

Würbel H, Chapman R, Rutland C 1998. Effect of feed and environmental enrichment on development of stereotypic wire-gnawing in laboratory mice. Applied Animal Behaviour Science 60, 69-81

William M.S. Russell is a zoologist who, with the late Rex L. Burch, when working for UFAW published in 1959 The Principles of Humane Experimental Technique, the pioneering book for animal welfare research that introduced the Three Rs: replacement of conscious animals by insentient material, reduction of numbers of animals used to obtain given information and refinement of procedures to minimize the distress imposed on the animals still used. Since 1990 William Russell has been Emeritus Professor at the Department of Sociology, University of Reading; he remains actively involved in laboratory animal science and welfare.

Comfortable Quarters for Mice in Research Institutions

Chris M. Sherwin

Centre for Behavioural Biology, Department of Clinical Veterinary Science, Langford House, University of Bristol, Bristol BS40 5DU, United Kingdom; email: chris.sherwin@bristol.ac.uk

Before describing how we can keep mice in more comfortable housing, it is worth briefly revisiting the reasons why this should be attempted. Each year many millions of mice are used throughout the world in research institutes. As part of this process, historically it has been the norm to breed and house mice under highly standardised conditions, aiming to reduce variability in responses and the data from subsequent research. This has meant that laboratory housing for mice is typically small, barren and monotonous. For some time, it has been questioned whether such housing systems compromise the welfare of the inhabitants. There is now convincing evidence that standard laboratory housing does indeed result in behavioural and physiological responses indicative of animal welfare compromises. Perhaps of greater importance is recent evidence that housing animals under such conditions affects the animals so fundamentally (Prior and Sachser, 1995; Prusky et al., 2000; Würbel, 2001) that concerns are being expressed about the validity of the data and its applicability to other circumstances. This calls into question the very reason for the animals being housed in these conditions in the first place.

There are compelling welfare and scientific reasons why we should house laboratory mice under conditions more suited to their own species-specific needs. These two factors are addressed separately below, with an emphasis on how they are inter-related.

et al., 1998). All these behaviours are thwarted by standard husbandry and housing. Housing systems should allow animals to perform most natural behaviours (e.g., "The Five Freedoms"; Farm Animal Welfare Council, 1997) to avoid compromises of welfare. In addition, if animals are prevented from performing behaviours for which they have a strong motivation, this can lead to suffering and adverse mental states such as frustration, depression and anxiety (Dawkins 1990; Duncan, 1992; Sherwin and Nicol, 1998). Certainly, conventional standard laboratory housing prevents many natural and highly motivated behaviours [e.g., nesting, tunnelling, extensive locomotion]. As a result, mice in laboratory conditions frequently exhibit so-called abnormal behaviours, for example stereotypies (Würbel et al., 1996; Nevison et al., 1999a), indicating that mice experience chronic frustration when placed in conventional, non-enriched cages (Sherwin, 2000). Furthermore, the sensory capabilities of mice have rarely been considered in laboratory housing and husbandry design. Mice have sensory modalities that are sometimes very disparate to humans [discussed below]. Our historical ignorance of these sensory capabilities means that standard housing generally does not take into account the perceptions of mice. This is potentially the equivalent to rearing animals under conditions of sensory deprivation or interference [e.g., olfactory "white noise"] with all the concomitant compromises in welfare (Cummins et al., 1977; van Praag et al., 2000).

Welfare Reasons for Providing Comfortable Quarters

Laboratory housing for mice has evolved from designs that were initially primarily concerned with economics, human convenience and extreme standardisation of the environment. This means that in current systems, the behavioural requirements of the animal are largely not catered for, other than the basics of feeding and drinking (Figures 1a & b). When given the opportunity, laboratory mice show a diverse behavioural repertoire: they seek a wide variety of foods, are very active physically, form complex social organisations, build tunnels and construct nests (Jennings

Scientific Reasons for Providing Comfortable Quarters

Animals reared in barren conditions are generally more sensitive to environmental perturbations or differences between laboratories. Therefore, when mice reared in conventional, barren cages are moved to a new laboratory, their behaviour might not be representative of "normal" responses. Moreover, there is growing evidence that the minimalistic environments of laboratory mice impose constraints on behaviour and brain development such that many studies using these animals may have little external validity, particularly in neuroscience studies.

Comfortable Quarters for Laboratory Animals Reinhardt V, Reinhardt A (eds), 6-17. Animal Welfare Institute, Washington, DC 20007

Studies may achieve good internal validity [i.e., reduced variation between animals in the same experiment in the same laboratory], but there might be increased variation between animals undergoing the same experiment in different laboratories (Crabbe et al., 1999). This diminished external validity calls into question the reason for keeping mice in conventional, barren cages. It can arise in three ways (Würbel, 2001):

1.Neurophysiological changes. It is well established that in rats, barren environments, compared to enriched ones, result in decreased numbers of brain neurones, synapses and dendritic branches, especially in the cortex and hippocampus. This results in impaired learning and memory (Rosenzweig and Bennett, 1996; van Praag et al., 2000). It has been argued that such effects suggest animals reared under standard laboratory conditions experience something akin to sensory deprivation (Cummins 1977; van Praag et al., 2000).

2.Chronic thwarting of behavioural response rules. Animals respond largely according to evolved rules that depend on specific environmental features. If these features are absent, the animal might display inappropriate behaviour. For example, gerbils will develop stereotyped digging if they do not have a suitable shelter. This stereotyped behaviour is prevented if they are presented with a shelter, but only if the shelter has a tunnel-shaped entrance (Wiedenmayer, 1997). Thus, barren cages can limit the opportunity for animals to develop appropriate behaviour with consequences for later studies in which it is assumed the animal is behaving "normally." In addition, chronic thwarting of behavioural responses can lead to stereotypies, associated with functional changes in the dorsal basal ganglia and a general tendency to perseverity [inability to be flexible in behaviour] (Albin et al., 1989; Ridley, 1994; Hauber, 1998). This could easily lead to scientists using an animal model that is fundamentally flawed.

3.Mismatch between postnatal and adult environment. The barren, monotonous environment in which mice are reared very often conflicts with the more variable life of mice undergoing research later in adult life. This can have a profound influence on the validity of the research. Laboratory-reared mouse pups are less stimulated than their wild counterparts as the laboratory-mother leaves the nest less frequently and for only short periods because food and water are provided nearby. However, if pups are handled for just a few minutes each day, the mother increases visits to the nests and grooms the pups to a degree that is thought to more accurately represent what is adaptive in the wild. As a consequence, pups who have been handled and thus better attended to by the mother [which might be encouraged by cage design and enrichment rather than handling] show reduced behavioural and endocrine responses to stress (Würbel, 2001). This could manifest as a reduced sensitivity to procedures causing pain, distress or suffering. In addition, barren environments at an early age can lead to improper development of the senses, e.g., vision (Prusky et al., 2000), with obvious consequences for studies requiring "normality" of these senses. The mismatch between postnatal and adult environment caused by standard laboratory housing is likely to have considerable, multifarious implications for research conducted on animals housed in these systems.

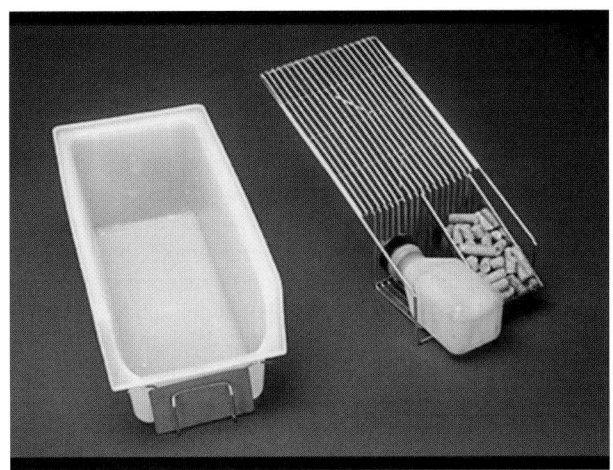

Figure 1a, b. A human perspective and a mouse perspective of a standard laboratory cage. The inside view of a standard cage shows this design caters little for the species-specific characteristics other than feeding and drinking.

There is often great concern about the harm, distress and suffering caused by particular research procedures upon mice, however, it must be remembered that almost ALL mice used in research—including breeders, stock, "spare" mice, and mice used in non-invasive studies—will be placed under standard housing conditions. This may be for considerable periods of time before the study itself and might persist afterwards. Therefore, the suffering caused by inappropriate housing and the lack of enrichment may well be of greater duration than the suffering caused by research-related activities. Arguably, the intensity of housing-related suffering might also be of greater intensity than that caused by the study. What is of great importance, however, is that the suffering caused by inappropriate housing and the lack of suitable environmental enrichment, will be experienced by a great proportion of all laboratory mice.

Providing Comfortable Quarters for Mice

To make life more comfortable for laboratory mice, we need to understand their species-specific characteristics. In particular, we need to understand their sensory perceptions and motivations. Therefore, the following sections summarise what we know about how mice perceive the world and how they behave. This is discussed in the context of how we might provide a better environment to address these species-specific requirements. It should be remembered, there is now available a bewildering diversity of strains of laboratory mice. Many of these have specific requirements [e.g., nude and ob-ob mice may have problems with thermoregulation]. Similarly, wild mice are different in many ways from their laboratory cousins (Jennings et al., 1998). It is impossible to consider all these requirements in this volume, so caretakers and investigators should make themselves aware of any particular idiosyncrasies of the strain they are working with and take appropriate action.

Comfortable Quarters and Sensory Perceptions of Laboratory Mice

Laboratory mice have [at least] the same five senses as humans, but these are used in different ways, which can make them difficult for us to imagine. Because we are unable to perceive the world in the same way as a mouse, and we have been historically ignorant of their senses, several aspects of the environment we provide have a direct impact on their perceptual capabilities and, ultimately, welfare status.

Olfaction

We humans use olfaction consciously very little, so this aspect of the mouse's perceptive world is almost totally hidden from us. But, olfaction is perhaps the most important sense used by mice, particularly in their highly complex social organisation. It is therefore important as a welfare concern, to understand the role of olfaction in mouse behaviour.

Mice create patterns of urine deposition for territorial marking and individual as well as group recognition (Hurst et al., 1993, 1998; Humphries et al., 1999; Nevison et al., 2000). Odours from adult males or from pregnant or lactating females can speed up or retard sexual maturation in juvenile females, and synchronise reproductive cycles in mature females. Odours of unfamiliar male mice may terminate pregnancies (Jennings et al., 1998). It has been shown that laboratory mice rendered surgically anosmic and then housed in large semi-natural enclosures interact with each other very differently from intact mice. Anosmic mice show very little aggression, roam freely about the enclosure rather than confining themselves to particular areas, and generally ignore each other. When they do encounter another individual, they appear startled and move away from each other (Liebenauer and Slotnick, 1996).

The use of olfaction by mice in mediating social encounters means that cage cleaning can be problematic (Gray and Hurst, 1995). There are two conflicting pressures: the need to clean cages for hygiene and health, and the need not to disturb scent-marking patterns too frequently. It has been shown that standard methods of cage cleaning, in which only the substrate and parts of the cage are washed clean of scent marks can be detrimental to male mice by promoting aggression. It has been recommended that if aggression is likely to be a problem, mice should be transferred into completely clean cages with fresh bedding (Jennings et al., 1998) or nesting material is moved with the mice (van Loo et al., 1988; Figure 2). It has also been suggested that strange odours [e.g., those associated with humans such as perfumes and deodorants] can produce stress responses in laboratory mice (Dhanjal, 1991). This should be taken into account when cleaning cages and handling the animals.

Clearly, olfaction plays a critical role in the social behaviour of mice. Therefore, it is of great concern that

Figure 2. Nesting material is easily provided for mice in standard laboratory cages. This is an inexpensive enrichment yet it probably represents the most cost-effective enrichment that can be given to mice in terms of its great impact on improving welfare and reducing aggression after cage cleaning.

inbreeding of laboratory mice can result in male mice becoming unable to discriminate between their own scent marks and those of other males (Nevison et al., 2000). This obviously could have considerable influence on behaviours such as agonistic and aggressive encounters and could easily affect responses in experiments. In addition, the lack of olfactory stimulation at an early age might influence performance in behavioural studies dependent on this sensory modality [e.g., discrimination studies] (Mihalick et al., 2000; Forestell, et al., 2001), learning and memory (Schellinck et al., 2001) and predator-related studies (Dellomo and Alleva, 1994).

Rats are natural predators of mice, and mice will show fear responses when they encounter anaesthetised rats (Blanchard et al., 1998). If housed in the same room, mice might become aware of rats by olfaction, even if they can not see them. Therefore, mice and rats should always be kept in separate rooms (Jennings et al., 1998). Similarly, it may be a wise precaution to change clothing and wash hands after handling predator species, such as rats and cats, or their bedding to avoid causing fear reactions in mice.

Hearing

Mice can hear over a broad spectrum of frequencies. They can detect frequencies from 80 Hz up to 100 kHz but are most sensitive in the 15 kHz to 20 kHz range and around 50 kHz (Jennings et al., 1998). This means they can hear well above the frequency of human hearing sensitivity.

Both audible and ultrasonic calls are used by mice in a variety of situations. Audible vocalisations can often be heard during agonistic encounters, whereas ultrasound is known to be used in sexual communication and also by pups when they have fallen out of the nest. It has even been reported that rats and shrews use ultrasound for echolocation (Kaltwasser and Schnitzler, 1981; Forsman and Malmquist, 1988).

Like many other laboratory mammals, mice are more sensitive than humans to sudden bursts of noises. They probably find sound pressure levels aversive when these are at an intensity 20 dB less than humans find aversive. Juvenile mice can become sensitised to loud sounds, including ultrasound. In some strains, this can increase the incidence of audiogenic convulsions or seizures (Gamble, 1982), decreased activity, reduced fertility and changes in blood glucose and corticosteroid levels [i.e., indicators of chronic stress].

Several items of common laboratory equipment such as pressure hoses, running taps, computer monitors or oscilloscopes can emit ultrasound at very high sound pressures (Sales et al., 1988,1999). This is silent to humans, but could have considerable effects on the welfare of mice. Laboratory-generated ultrasound may interfere with communication between mice, causing distress or perhaps even sensory damage. This is likely to remain undetected by the staff. Furthermore, when many mice are housed in a large laboratory, or in high stocking densities within cages, the number of ultrasonic calls being given at any one time could be very great. This could result in the mice perceiving the environment to be very noisy and potentially stressful—rather like humans being in a crowded room with everyone shouting to each other—but human animal attendants would be totally unaware of this noise. Commercial "bat detectors" register ultrasonic frequencies and could be used to detect whether equipment is generating ultrasound, or when mice are vocalising to each other or in response to distressful procedures such as blood collection and injection. In this way, monitoring of ultrasonic vocalisations could be used to help in overall assessment of mouse welfare.

Items that routinely make loud noises [e.g., alarms, telephones, door-bells] should be designed to operate at frequencies less audible to the rodent ear [e.g., below 500 Hz]. It has often been suggested that a radio playing softly in the background can be a suitable enrichment and that this makes mice more tractable and less responsive to sudden noises. I have not seen any scientific evidence to support this, and it should be remembered that at least some of the sound produced by radios will be below the frequencies to which mice are sensitive. However, the radio may provide more comfortable quarters for the human caretakers, which in turn could have beneficial consequences for the animals.

Vision

Although mice have good vision, this sense is perhaps less important than others. In the wild, most mice are nocturnal and usually avoid brightly lit areas. Therefore, the light intensities in which we keep laboratory mice are relatively high compared to the environment in which they have evolved. These higher light intensities can lead to eye abnormalities, including the induction or exacerbation of retinal atrophy (Jennings et al., 1998). This, of course, can result in gross disturbances to visual perception, a particular problem for albino strains lacking protective visual pigment, and for individuals in top-row cages that are not shaded by a row of cages above them. To protect mice from intense illumination, cages—especially transparent cages—should be provisioned with nesting material, and a light baffle [sheet of wood or plastic] placed over the top row.

Prusky et al. (2000) demonstrated that enriching the environment of mice early in life led to significantly improved vision. Pups reared from birth in large, clear cages with enrichment objects had 18% better acuity than pups reared under standard laboratory conditions. This shows clearly that the mouse's visual system can be significantly influenced by the nature of early visual input.

The visual apparatus of mice is basically similar to humans, with the exception of its sensitivity to ultra-violet light. The ventral area of the mouse retina has a denser accumulation of ultra-violet sensitive cones (Calderbone and Jacobs, 1995; Szel et al., 1996; Yokohyama and Shi, 2000; Neitz and Neitz, 2001). The biological significance of this structure is not yet known. Humans with normal vision are insensitive to ultra-violet light [the cornea blocks it] and as a result of this, we have designed artificial lights, including those used in laboratories, to emit very little ultra-violet radiation. Although the consequences of housing mice under lights with little ultra-violet output have not been experimentally determined, other species with ultra-violet sensitivity prefer areas supplemented with these wavelengths (Moinard and Sherwin, 1999) and housing them without ultra-violet causes physiological and behavioural disturbances indicative of welfare compromise (Maddocks et al., 2001). Placing an animal with ultra-violet sensitivity into an environment without these wavelengths is likely to distort their perception of <u>all</u> colours, not simply seeing the world minus one colour. These colour shifts could give the mouse a distorted visual perception of his/her world, rather like humans seeing a psychedelic picture. Although the welfare implications of ultra-violet sensitivity are undetermined, this should be considered and further research in the area encouraged.

The above studies all indicate that the visual environment of standard laboratory housing is often inappropriate for mice and can lead to impaired vision. Many behavioural tests in research institutes are totally dependent on vision, yet we rarely account for the visual experience of the animal and how "normal" his or her perceptive capabilities might be. Also related to vision and behaviour, mice are generally nocturnal, yet routine husbandry—including handling and inspection of the animals—and many tests are performed under bright lights or during the light phase when the animal would normally be asleep. This leads one to question, perhaps rather anthropomorphically, the validity of data from such tests: would we trust data from behavioural and physiological tests when these had been gathered from humans chronically sleep-deprived by being repeatedly woken up in the middle of the night!

Whenever possible, the lights should be put onto a reverse light:dark cycle. With the aid of a dim red light, the mice can be inspected and handled during the dark phase when they would normally be active. In addition, lights can be programmed to gradually increase or decrease in intensity to provide an artificial dawn or dusk, a circumstance that is thought to give animals the opportunity to prepare for periods of inactivity or activity [e.g., by eating a little extra food, or building nests].

There is some evidence that mice prefer opaque cages to transparent cages (Baumans et al., 1987). This might be related to light intensity, or seeing neighbouring mice in close proximity but being unable to assess their status by olfaction. Whichever, providing nesting material will benefit the animals if they have to be kept in transparent cages.

Touch

Touch is an important sense for mice. In times of stress, mice retain contact with surfaces (Berry, 1981). Loss of tactile contact with conspecifics seems to be the most important factor in determining the isolation-induced increase in aggressiveness in male mice (Brain and Benton, 1983). When moving about, mice like to remain in contact with a wall and generally show avoidance of open spaces, i.e., thigmotaxis. This preference can be catered for in cages by providing dividers in either the vertical or horizontal planes. They add complexity to the cage and might also increase the size of the cage as perceived by the mice. It has been reported that vertical cage dividers can reduce fearful or anxious behaviour in a novel environment (Boyd and Love, 1995). Some commercial pet companies produce "mazes" for rodents that could be practical enrichment for laboratory mice. The "mazes" are essentially small cages with many vertical dividers and a transparent, removable lid making it easy to locate and catch the mouse.

The facial-vibrissae, or whiskers, are acutely sensitive touch organs used to investigate novel objects, or during thigmotaxis when moving about the environment. Laboratory mice sometimes engage in a behaviour called whisker-trimming, or barbering, in which one mouse trims the vibrissae of another, sometimes totally. The significance of this behaviour is unclear. It has been hypothesized to be a dominance behaviour (but see van de Weerd 1992; Garner et al., 2001), however, rather bizarrely, it has been shown that if a pair of mice are separated by wire mesh, whisker-trimming continues (Vandenbroek et al., 1993). This indicates some co-operation by the mouse who is being whisker-trimmed, and it was argued that whisker-trimming might lead to the release of endorphins [i.e., mice co-operate in this behaviour as a form of "coping response" to a stressful situation]. Whisker-trimming does not appear to have been reported in wild mice, suggesting that laboratory conditions might predispose this "abnormal behaviour." Whatever the underlying cause or function, if laboratory conditions result in whisker-trimming, this indicates the standard environment potentiates a behaviour that causes at least some mice to lose one aspect of their sense of touch. Therefore,

Figure 3. A standard mouse cage enriched with a pet rodent "drop feeder" to make the mice work for an alternative food, shredded paper for nesting material, a running wheel and a chewing block.

any enrichment that prevents this behaviour is probably improving welfare.

Many mouse cages in research institutes have wire mesh or grid floors. These prevent us from providing mice with floor substrate. This in turn thwarts several behaviours that the mice are motivated to perform and can also result in health problems such as pressure sores (Hubrecht, 1995) and urological problems (Everitt et al., 1988). These problems are addressed by the *Guide for the Care and Use of Laboratory Animals*, stipulating that "Solid-bottom caging, with bedding, is therefore recommended for rodents" (National Research Council, 1996, p. 24). Blom (1993) showed that mice prefer a solid resting site and will generally spend more time on a solid surface rather than a grid floor (Blom et al., 1996). Although it might not be possible to change the entire floor structure, at least a section of the cage floor should be covered, or a receptacle of substrate provided for resting. The mice will avoid soiling this substrate [see "Eliminative behaviour"], meaning it stays relatively clean and does not have to be changed frequently.

Taste

Mice are used as a model species for humans in studies relating to taste. Presumably then, mice have taste apparatus and taste sensations similar to those of humans. Wild mice will eat a wide range of foods such as seeds, fresh vegetables, fruit and bread (Jennings et al. 1998), and where this does not interfere with the objectives of the research, such food should be offered as an alternative to or a supplement of the standard laboratory rodent pellet. If a standard diet is given, expanded forms appear to be more palatable than pelleted, presumably due to the difference in texture and taste (Jennings et al., 1998).

Differences in taste preferences have been reported for different strains of laboratory mice (Frank and Blizard, 1999). In addition, behavioural studies have shown that mouse pups readily develop preferences for the same food that their mother eats, and that the strength of this preference is dependent upon the taste properties of the food (Valsecchi et al., 1993).

Food can affect odour cues of mice, and a group of animals kept on the same diet may have more difficulties discriminating between individual group members than animals kept on a variety of different diets (Brown et al., 1996). This suggests that providing varied diets might increase the ability of cage-mates to discriminate amongst themselves and therefore possibly reduce agonistic or aggressive encounters. Scalera (1992) observed in rats that taste preferences, water consumption and food consumption can all be significantly different depending on whether the animals are housed singly, in pairs or groups and warned that for experiments in which appetite and taste are dependent variables, the animals should be housed under similar social and environmental conditions.

A commercial product [rather comically designed as a huge mouse] has recently become available for pet rodents (Figure 3). Favoured food is placed in the gadget, which has several holes. The aim is to encourage the mouse to push the puzzle about the cage to get the food to eventually drop through one of the holes. Although the effectiveness

Figure 4. A cup-shaped nest. In this study, the mice removed both types of bedding material offered to them in the containers and built a composite nest.

of this gadget as an enrichment tool for laboratory mice has not been tested, the principles of increasing food diversity and allowing the animal to work for food suggest this would improve welfare.

Comfortable Quarters and Species-specific Behaviours of the Laboratory Mouse

The behaviour of laboratory mice can be complex. Detailed ethological analyses of caged animals reveal more than 40 different, commonly exhibited activities and postures (Jennings et al., 1998). However, there are several behaviours that laboratory mice readily perform when given the opportunity but which are normally thwarted by the small size or barrenness of standard cages. These are outlined below, along with suggestions of how enrichment might allow mice the opportunity to perform these behaviours.

Nest-building

Mice will build nests with much apparent enthusiasm (Figure 2). This not only helps them to protect their young,

Figure 5. Running wheels are suitable to promote exercise in captive mice.

but in non-breeding animals can also help to regulate temperature and light levels and to hide and retreat from cage-mates or other threatening stimuli. There is considerable evidence that mice are strongly motivated to build nests (Blom 1993; Sherwin 1996, 1997; van de Weerd et al., 1998), indicating that this activity fulfils one of their most fundamental behavioural needs.

Several products are commercially available, but hay, straw, shredded paper, wood chips and paper tissues are all useful. Paper towels can be left on the cage lid for the mice to energetically drag through the bars and chew into pieces to build a nest. If several materials are available, mice will generally build a composite nest (Figure 4). Although providing nesting material is easily and inexpensively achieved, it is one of the best and most versatile enrichments for mice kept in research institutions. Suitable nesting supply should be provided for all animals, although for cages with new-born young, materials that might entrap legs should be avoided [e.g., cotton wool, wood wool, shredded paper] and materials that absorb moisture as these can stick to wet pups and cause dehydration.

Increased Activity

Mice are extremely active animals, yet the physical dimensions of a standard sized cage allow mice to move only a few body lengths in any one direction. This spatial restriction, in conjunction with a plentiful supply of food nearby, means that mice can quickly become overweight with a subsequent reduction in life-span. They demonstrate a strong motivation to gain access to additional space to that provided by a standard laboratory cage, even when this provides no further resources or enrichment (Sherwin and Nicol, 1997); this could be interpreted as the mice having a strong urge to escape standard laboratory conditions!

One method of providing an opportunity for increased activity is a running wheel (Figure 5). There is much evidence to suggest that providing a running wheel is of great benefit. Mice will work hard to gain access to a wheel, they prefer a wheel to an extended surrogate tunnel system, and there are many physiological and behavioural advantages related to welfare. Mice sometimes appear to play in running wheels. For example, they will grip the rungs of the wheel until they are carried around and around by the wheel's momentum. They will turn motorised wheels on and off. It has even been reported that mice prefer wheels that have been made into irregular shapes, or include hurdles to jump over (Sherwin, 1998 a,b).

Other methods can be utilised to encourage increased activity, even within the confines of a relatively small cage. Simple activity discs can be made relatively easily and cheaply (Figure 6; Animal Welfare Institute, 1979). Commercial pet companies manufacture "activity dishes," which resemble a miniature satellite dish set at an angle to rotate about a central axis (Figure 9). Climbing frames, ropes, pieces of string or chains all allow mice to climb. In addition,

Figure 6. Activity discs can be made relatively easily and cheaply (photo by Ernest P. Walker, 1979).

the bars of the cage-lid are used prodigiously; if taller cages are used, enrichments allowing access to the lid should be provided. For this reason, amongst others, cages with solid tops are not recommended.

Tunnel-building
Many wild rodents build complex tunnel systems (Ellison, 1993; Schmid-Holmes, 2001). These are used to escape predators (Blanchard et al., 1995) and presumably for other comfort factors including thigmotaxis. Laboratory mice who have never had the opportunity to dig tunnels will build these within a few hours if a suitable substrate is provided (Sherwin, personal observation; Figure 7). Unfortunately, providing mice with the opportunity for tunnelling can make them rather difficult to catch although they will often sleep in an attached cage that leads to the tunnelling substrate. However, if regular handling is not required, or naturalistic behaviour is desirable, several centimetre deep, suitable substrate [e.g., damped peat with rocks or fibrous bedding to support the tunnel system] provides for almost instantaneous digging and some very entertaining mouse behaviour. Wood chip bedding might be a suitable compromise as it allows mice to perform digging behaviour and seek a darker environment but does not allow them to totally escape detection from human concerns. Surrogate burrows can be offered in the form of plastic tubes designed for pet rodents; several types are available commercially. Laboratory mice seem to gain a great sense of security in these tunnels even when they are transparent; they often appear completely oblivious to nearby human presence. Providing tubes as tunnels can also make catching the mice a little difficult, although the tunnels can usually be separated into smaller sections and the one containing the mouse placed into the cage he or she is being transferred to; the mouse then usually walks out of the tube within a few seconds. Alternatively, if there are short tubes, mice use these as retreats and run into them during attempted capture. The tube containing the mouse is then easily transferred elsewhere, or the protruding tail of the mouse used for quick and easy handling, which also reduces stress caused to the mouse.

As described before, commercial pet companies produce "mazes" for pet rodents that might provide a practical surrogate tunnel system for laboratory mice in some situations.

Chewing/gnawing
Mice will readily chew on a variety of objects and should be provided with the opportunity to express this behaviour. Such chewable objects might include cardboard tubes, softwood blocks, old plastic water bottles, hay, straw, etc. (Figure 8). Cardboard tubes are particularly versatile as they also provide opportunities for shelter, climbing and manipulation.

Thermoregulation
Some rodents prefer cooler ambient temperatures in the dark phase and warmer temperatures during the light phase (Gordon, 1993). This suggests that a diurnally changing temperature might contribute to improving the animals'

Figure 7. Laboratory mice who have never encountered deep substrate will readily dig tunnels when given the opportunity.

Figure 8. This enrichment is advertised as a wooden chewing block, but its design allows it to also be used as a nest/shelter (the mice drag paper into it) and a climbing object.

comfort. Of course, providing suitable nesting material is likely to circumvent the preference for changing ambient temperature and also provides the opportunity for other

Figure 9. Mice will sometimes defecate in highly localised areas. This behaviour can be promoted by giving the mice a demarcated area, such as the petri-dish in the corner of this cage. The other object is an activity dish.

behaviours. Where metal cages are used, the "coldness" of this material can be overcome at least partly by provision of much bedding and nesting material.

Eliminative behaviour

Laboratory mice will often deposit their faeces in specific sites or latrines (Sherwin, 1996b; Blom, 1993; Figure 9). This behaviour could be involved in signals for social communication, hiding from potential predators, or it could simply be a hygiene response. Whichever, the small, featureless environment of a standard cage gives a mouse little choice to select certain areas or to avoid those marked by other individuals. Providing objects that are easily demarcated, such as vertical dividers, tins, jars, etc., will allow mice to show this eliminative behaviour pattern.

Social behaviour

Mice are a highly social species and, where possible, should be maintained in stable, harmonious groups. There are sound scientific and welfare reasons for this. Individual housing can influence responses to laboratory procedures (Mackintosh, 1962), and mice generally show clear preferences to be in close proximity to other mice, even males who have been housed singly (Vandenbroek et al., 1993; van Loo et al., 2001).

Forming groups: It seems a good general principle to form groups from weanlings who know each other. Same-sex groups are best set up before the animals reach the age of puberty. The likelihood of aggressive incompatibility increases in older animals, especially unfamiliar males (Barnard et al., 1991). When forming groups, various factors must be considered such as sex, age, reproductive condition, etc. Groups should be established in clean cages, as home-cage odour cues induce residents to attack intruders, and unfamiliar odours can increase aggression amongst the residents (Brown, 1985). It is generally not good practice to take mice from non-enriched cages and form new groups in enriched cages as this can result in territorial, aggressive disputes (McGregor and Ayling, 1990).

Maintenance of groups: Mice form a complex social organisation, and each animal plays a role in it, sometimes dependent on factors including age, sex, position in the hierarchy, or reproductive condition (Jennings et al., 1998). Upsetting this organisation by addition or removal of individuals, perhaps only one, can have considerable consequences that might ultimately affect the welfare of <u>all</u> mice within the group. There is generally no problem in group-housing young mice and non-breeding females. Housing mature males together is more of a problem, especially in smaller groups of two or three, which can put excessive amounts of stress on the subordinate animal(s). Aggression levels depend on a great many factors such as age, group size, cage size, previous experience of the animals and the situation (Jennings et al., 1998). The provision of cornhusk can buffer aggressive tension by offering subordinate animals cover and escape routes (Armstrong et al., 1998). Dominant males tend to be more aggressive in an environment with familiar odour than in a strange environment. Complete cage cleaning—new cage and new substrate—can, therefore, minimise aggression among male mice compared to partial cage cleaning (Gray and Hurst, 1995). Van Loo et al., (2000) observed that transferring nesting material during cage cleaning reduced aggression among males, whereas transferring sawdust containing urine and faeces seemed to intensify aggression.

It has been reported that providing enrichments or cage "furniture" for group-housed mice can increase aggression (McGregor and Ayling 1990; Haemisch et al., 1994). It was noted in one of these papers (McGregor and Ayling, 1990) that the mice might easily have regarded these objects as resources and thus defended them. Whilst trying to avoid being dismissive of these papers, the authors have assumed that sufficient space and refuges were provided for subordinate animals to show appeasement behaviour, escape, etc. Whenever enrichments are offered to mice, these should be in sufficient number and at a sufficient distance so that aggressive competition is not triggered.

Final Comments

In providing comfortable quarters for mice, there are several "-isms" we should avoid. We should avoid speciesism and remember there is no evidence to suggest that mice do NOT have the same capacity to suffer as other vertebrates, although their suffering might occur in different ways. We should also avoid anthropomorphism and anthropocentrism, and try to understand the mouse's world from it's own perspective, rather than our own human concerns. We should also avoid sizeism: simply because laboratory mice are small and can all appear to be the same [at least to us], this does not mean they have any less capacity to suffer as individuals.

In promoting appropriate housing for mice, it can be helpful to think in terms of optimising the 2 Qs: Quantity of space and Quality of space. We should be aiming to provide mice with the appropriate amount of space containing the appropriate diversity of environment that takes into account their species-specific characteristics and needs.

Acknowledgements

C.M. Sherwin was funded by the UFAW Hume Research Fellowship during preparation of this chapter.

References

Albin RL, Young AB, Penney JB 1989. The functional anatomy of basal ganglia disorders. Trends in Neurosciences 12, 366-375

Animal Welfare Institute 1979. Comfortable Quarters for Laboratory Animals, Seventh Edition. Animal Welfare Institute, Washington, DC

Armstrong KR, Clark TR, Peterson MR 1998. Use of cornhusk nesting material to reduce aggression in caged mice. Contemporary Topics in Laboratory Animal Science 37(4), 64-66

Barnard CJ, Hurst JL, Aldhous P 1991. Of mice and kin: the functional significance of kin bias in social behaviour. Biological Reviews 66, 379-430

Baumans V, Stafleu FR, Bouw J 1987. Testing housing systems for mice—the value of a preference test. Zeitschrift für Versuchstierkunde 29, 9-14

Berry RJ (ed) 1981. Biology of the House Mouse. In Symposium of the Zoological Society of London, No 47. Academic Press, London, UK

Blanchard RJ, Hebert MA, Ferrari P, Palanza P, Figueira R, Blanchard DC, Pamigiani S 1998. Defensive behaviour in wild and laboratory (Swiss) mice: the mouse defence test battery. Physiology and Behavior 65, 201-209

Blom HJM 1993. Evaluation of Housing Conditions for Laboratory Mice and Rats. The Use of Preference Tests for Studying Choice Behaviour. Utrecht University, Utrecht, Netherlands

Blom HJM, Van Tintelen G, Van Vorstenbosch CJAVH, Baumans V, Benyen AC 1996. Preferences of mice and rats for type of bedding material. Laboratory Animals 30, 234-244

Boyd J, Love JA 1995. The effects of dividers on the nesting sites of mice. Frontiers in Laboratory Science, oral presentation. Helsinki, Finland

Brain PF, Benton D 1983. Conditions of housing, hormones and aggressive behaviour. In Hormones and Aggressive Behaviour Svare BB (ed), 349-372. Plenum Press, New York, NY

Brown RE 1985. The rodents II: suborder Myomorpha. In Social Odours in Mammals, Volume 1 Brown RE, MacDonald DW (eds), 345-457. Clarendon Press, Oxford, UK

Brown RE, Schellink HM, West AM 1996. The influence of dietary and genetic cues on the ability of rats to discriminate between the urinary odors of MHC-congenic mice. Physiology and Behavior 60, 365-372

Calderbone JB, Jacobs GH 1995. Regional variations in the relative sensitivity to UV light in the mouse retina. Visual Neuroscience 12, 463-468

Crabbe JC, Wahlsten D, Dudek BC 1999. Genetics of mouse behavior: Interactions with laboratory environment. Science 284, 1670-1672

Cummins RA, Livesey PJ, Evans JGM 1977. A developmental theory of environmental enrichment. Science 197, 692-694

Dawkins MS 1990. From an animal's point of view: Motivation, fitness, and animal welfare. Behavioral and Brain Sciences 13, 1-61

Dellomo G, Alleva E 1994. Snake odour alters behaviour, but not pain sensitivity in mice. Physiology and Behavior 55, 125-128

Dhanjal P 1991. The Assessment of Stress in Laboratory Mice Due to Olfactory Stimulation with Fragranced Odours. M.Sc. Dissertation, University of Birmingham, Birmingham, UK

Duncan IJH 1992. Designing environments for animals—not for public perceptions. British Veterinary Journal 148, 475-477

Ellison GTH 1993. Group-size, burrow structure and hoarding activity of pouched mice in Southern Africa. African Journal of Ecology 31, 135-155

Everitt JI, Ross PW, Davis TW 1988. Urologic syndrome associated with wire caging in AKR mice. Laboratory Animal Science 38, 609-611

Farm Animal Welfare Council 1997. Report on the Welfare of Laying Hens. Farm Animal Welfare Council, Tolworth, UK

Forestell CA, Schellinck HM, Boudreau SE, LoLordo VM 2001. Effect of food restriction on acquisition and expression of a conditioned odor discrimination in mice. Physiology and Behavior 72, 559-566

Frank ME, Blizard DA 1999. Chorda tympani responses in two inbred strains of mice with different taste preferences. Physiology and Behaviour 67, 287-297

Forsman KA, Malmquist MG 1988. Evidence for echolocation in the common shrew. Journal of Zoology 216, 655-662

Gamble MR 1982. Sound and its significance for laboratory animals. Biological Reviews 57, 395-421

Garner JP, Weisker SM, Dufour B, Gregg LE, Mench JA 2001. The epidemiology of barbering (whisker trimming) in laboratory mice. Proceedings of the 35th International Society of Applied Ethology International Congress, 129

Gordon CJ 1993. Twenty-four hour rhythms of selected ambient temperature in rat and hamster. Physiology and Behavior 53, 257-263

Gray S, Hurst JL 1995. The effects of cage cleaning on aggression within groups of male laboratory mice. Animal Behaviour 49, 821-826

Haemisch A., Voss T, Gärtner K 1994. Effects of environmental enrichment on aggressive behavior, dominance hierarchies, and endocrine status in male DBA/2J mice. Physiology and Behavior 56, 1041-1048

Hauber W 1998. Involvement of basal ganglia transmitter systems in motor initiation. Progress in Neurobiology 56, 507-540

Hubrecht R 1995. Housing, Husbandry and Welfare Provision for Animals used in Toxicology Studies: Results of a UK Questionnaire on Current Practice (1994). Universities Federation for Animal Welfare, Potters Bar, UK

Humphries RE, Robertson DHL, Beynon RJ, Hurst JL 1999. Unravelling the chemical basis of competitive scent marking in house mice. Animal Behaviour 58, 1177-1190

Hurst JL, Fang J, Barnard CJ 1993. The role of substrate odours in maintaining social tolerance between male house mice. Animal Behaviour 45, 997-1006

Hurst JL, Robertson DHL, Tolladay U, Beynon RJ 1998. Proteins in urine scent marks of male house mice extend the longevity of olfactory signals. Animal Behaviour 55, 1289-1297

Jennings M, Batchelor GR, Brain PF, Dick A, Elliot H, Francis RJ, Hubrecht RC, Hurst JL, Morton DB, Peters AG, Raymond R, Sales GD, Sherwin CM, West C 1998. Refining rodent husbandry: the mouse. Report of the Rodent Refinement Working Party. Laboratory Animals 32, 233-259

Kaltwasser MT, Schnitzler HU 1981. Echolocation signals confirmed in rats. Zeitschrift für Säugetierkunde 46, 394-395

Liebenauer LL, Slotnick BM 1996. Social organisation and aggression in a group of olfactory bulbectomized male mice. Physiology and Behavior 60, 403-409

Mackintosh JH 1962. Effect of strain and group size on the response of mice to "sconal" anaesthesia. Nature 194, 1304

Maddocks SA, Cuthill IC, Goldsmith AR, Sherwin CM 2001. Behavioural and physiological effects of absence of ultraviolet wavelengths for domestic chicks. Animal Behaviour 62, 1013-1019

McGregor PK, Ayling SJ 1990. Varied cages result in more aggression in male CFLP mice. Applied Animal Behaviour Science 26, 277-281

Mihalick SM, Langlois JC, Krienke JD, Dube WV 2000. An olfactory discrimination procedure for mice. Journal of the Experimental Analysis of Behavior 73, 305-318

Moinard C, Sherwin CM 1999. Turkeys prefer fluorescent light with supplementary ultraviolet radiation. Applied Animal Behaviour Science 64, 261-267

National Research Council 1996. Guide for the Care and Use of Laboratory Animals, 7th Edition. National Academy Press, Washington, DC
Full Text: http://www.nap.edu/readingroom/books/labrats/

Neitz M, Neitz J 2001. The uncommon retina of the common house mouse. Trends in Neurosciences 24, 248-249

Nevison CM, Hurst JL, Barnard CJ 1999. Why do male ICR (CD-1) mice perform bar-related (stereotypic) behaviour? Behavioural Processes 47, 95-111

Nevison CM, Barnard CJ, Beynon RJ, Hurst JL 2000. The consequences of inbreeding for recognising competitors. Proceedings of the Royal Society of London, Series B 267, 687-694

Prior H, Sachser N 1995. Effects of enriched housing environment on the behaviour of young male and female mice in four exploratory tasks. Journal of Experimental Animal Science 37, 57-68

Prusky GT, Reidel C, Douglas RM 2000. Environmental enrichment from birth enhances visual acuity but not place learning in mice. Behavioural Brain Research 114, 11-15

Ridley RM 1994. The psychology of perseverative and stereotyped behaviour. Progress in Neurobiology 44, 221-231

Rosenzweig MR, Bennett EL 1996. Psychobiology of plasticity: effects of training and experience on brain and behavior. Behavioural Brain Research 78, 57-65

Sales GD, Wilson KJ, Spencer KE, Milligan SR 1988. Environmental ultrasound in laboratories and animal houses: a possible cause for concern in the welfare and use of laboratory animals. Laboratory Animals 22, 369-375

Sales GD, Milligan SR, Khirnykh K 1999. Sources of sound in the laboratory animal environment: a survey of the sounds produced by procedures and equipment. Animal Welfare 8, 97-115

Scalera G 1992. Taste preferences, body-weight gain, food and fluid intake in singly, or group-housed rats. Physiology and Behaviour 52, 935-943

Schellinck HM, Forestell CA, LoLordo VM 2001. A simple and reliable test of olfactory learning and memory in mice. Chemical Senses 26, 663-672

Schmid-Holmes S, Drickamer LC, Robinson AS, Gillie LL 2001. Burrows and burrow-cleaning behaviour of house mice. American Midland Naturalist 146, 53-62

Sherwin CM 1996a. Preferences of individually housed TO strain laboratory mice for loose substrate or tubes for sleeping. Laboratory Animals 30, 245-251

Sherwin CM 1996b. Preferences of laboratory mice for characteristics of soiling sites. Animal Welfare 5, 283-288

Sherwin CM, Nicol CJ 1997. Behavioural demand functions of caged laboratory mice for additional space. Animal Behaviour 53, 67-74

Sherwin CM 1997. Observations on the prevalence of nest-building in non-breeding, TO strain mice and their use of two nesting materials. Laboratory Animals 31, 125-132

Sherwin CM 1998a. The use and perceived importance of three resources which provide caged laboratory mice the opportunity of extended locomotion. Applied Animal Behaviour Science 55, 353-367

Sherwin CM 1998b. Voluntary wheel-running: a review and novel interpretation. Animal Behaviour 56, 11-27

Sherwin CM, Nicol CJ 1998. A demanding task: using economic techniques to assess animal priorities, a reply to Mason et al. Animal Behaviour 55, 1079-1081

Sherwin CM 2000. Frustration in laboratory mice. Scientists Centre for Animal Welfare Newsletter 22(3), 7-12

Szel A, Rohlich P, Caffe AR, van Veen T 1996. Distribution of cone receptors in the mammalian retina. Microscopy Research and Technique 35, 445-462

Valsecchi P, Moles A, Mainardi M 1993. Transfer of food preferences in mice at weaning—the role of maternal diet. Bollettino Di Zoologia 60, 297-300

van de Weerd HA, Vandenbroek FAR, Beynen AC 1992. Removal of the vibrissae in male mice does not influence social dominance. Behavioural Processes 27, 205-208

van de Weerd HA, Van Loo PLP, Van Zutphen LFM, Koolhaas JM, Baumans V 1998. Strength of preference for nesting material as environmental enrichment for laboratory mice. Applied Animal Behaviour Science 55, 169-382

Vandenbroek FAR, Omtzight CM, Beynen AC 1993. Whisker trimming behaviour in A2G mice is not prevented by offering means of withdrawal from it. Laboratory Animals 27, 270-272

van Loo PLP, Kruitwagen CLJJ, Van Zutphen LFM 2000. Modulation of aggression in male mice: Influence of cage cleaning regime and scent marks. Animal Welfare 9, 281-295

van Loo PLP, de Groot AC, Van Zutphen BFM, Bauman V 2001. Do male mice prefer or avoid each other's company? Influence of hierarchy, kinship, and familiarity. Journal of Applied Animal Welfare Science 4, 91-103

van Praag H, Kempermann G, Gage FH 2000. Neuronal consequences of environmental enrichment. Nature Reviews Neuroscience 1, 191-198

Wiedenmayer C 1997. Causation of the ontogenetic development of stereotypic digging in gerbils. Animal Behaviour 53, 461-470

Würbel H, Stauffacher M and vonHolst D 1996. Stereotypies in laboratory mice: quantitative and qualitative description of the ontogeny of "wire-gnawing" and "jumping" in ICR and ICR-nu mice. Ethology 102, 371-385

Würbel H 2001. Ideal homes? Housing effects on rodent brain and behaviour. Trends in Neuroscience 24, 207-211

Yokohyama S, Shi YS 2000. Genetics and evolution of ultra-violet vision in vertebrates. FEBS [Federation of European Biochemical Societies] Letters 486, 167-172

Dr. Chris Sherwin gained his Ph.D. at Murdoch University (Australia) in 1987 and after a period working in New South Wales, he moved to the University of Bristol (UK) in 1990. Since then, he has worked on a variety of subjects relating to animal behaviour and welfare. These include applied investigations of improved housing for layer hens, laboratory mice and turkeys, and more fundamental studies on social learning, motivation and preference tests. He is currently the UFAW Research Fellow in Animal Welfare and is investigating designs of cages to improve the welfare of laboratory mice.

Comfortable Quarters for Gerbils in Research Institutions

Eva Waiblinger

Animal Behavior, Zoological Institute, University of Zürich, Winterthurerstraße 190, CH-8057 Zürich, Switzerland; email: eviwai@zool.unizh.ch

Why is the Well-being of Gerbils so Important for Laboratories and Experimenters?

Most animal welfare laws and guidelines for the humane care of laboratory animals require:
(a) that the environment of captive animals meets their physiological and behavioral needs;
(b) that housing conditions facilitate the performance of natural behavior patterns;
(c) that husbandry conditions allow for adequate social contacts.

Hence, comfortable quarters for gerbils must be based on the species' physiological and behavioral needs, behavior repertoire and regulation mechanisms as well as social structure.

Housing-induced changes in brain and behavior

Behavior control mechanisms evolved in a species' natural environment and require a specific set of environmental and social stimuli to function properly (Hughes and Duncan, 1988). Gerbil breeders concentrated not on altering behavior mechanisms, but rather on changing specific behavior traits [tameness, reduced aggressiveness] and physiological parameters [higher reproductive output, higher body weight] compared to wild-caught gerbils (Stuermer, 1998). Despite decades of domestication, most behavioral traits and behavioral regulation mechanisms have been retained in laboratory gerbils (Wiedenmayer, 1995) as have their behavioral needs. However, the barren laboratory cage lacks crucial environmental and social stimuli involved in eliciting and regulating behavior as well as stimulating behavior and brain development. As a result, behavioral expressions become chronically thwarted and the animal's ability to adapt to these conditions is overtaxed (Mason, 1991; Würbel, 2001). Rearing rodents in such impoverished laboratory cages can result in impaired cognitive functions, inappropriate and ill-regulated behavioral responses, stress reactions, abnormal behaviors such as stereotypic digging and bar chewing, altered brain functions and reduced stress-tolerance in adulthood, which in consequence all threaten the validity of experimental results gained from such animals (Wiedenmayer, 1995; Winterfeld et al. 1998; Würbel, 2001). Housing and husbandry should, therefore, be adapted to presenting the animals with an appropriate set of stimuli to regulate behavior and forestall the development of stereotypies.

Standardization, variability and result validity

Standardization of experimental procedures requires a reduction of inter- and intra-individual variation (Würbel, 2000). Notwithstanding, very large individual variation of stereotypic behavior patterns can be observed in animals under barren housing conditions, which reflect their different strategies to cope with the situation (Ödberg, 1989; Dantzer, 1991; Mason, 1991; Wechsler, 1995). Animals kept under enriched conditions should yield more robust results and greater validity, since an enriched environment better resembles the stimulus-rich, variable conditions of natural environments (Würbel, 2000, 2001).

Comfortable Quarters for Laboratory Animals Reinhardt V, Reinhardt A (eds), 18-25. Animal Welfare Institute, Washington, DC 20007

What is environmental enrichment?

Environmental enrichment are those measures that add extra, species-appropriate environmental stimuli to standard laboratory housing conditions [e.g., 2-3 cm of bedding, pelleted food in food hopper, water ad libitum]. Environmental enrichment is often done intuitively and is based on availability and practicability of materials rather than on the species-specific needs of the confined animal. For enrichment to be most effective in eliciting natural behavior patterns and reducing the development of stereotypies, it should be designed on the basis of a thorough knowledge of the behavioral needs and behavior control mechanisms of the species (Wiedenmayer, 1996). Ethological studies under seminatural and enriched as well as under impoverished conditions can assess the influence of different environmental factors and stimuli on behavior and their relative importance for the well-being of the animals. Ideally, such studies should be accompanied by physiological measurements of welfare-indicators and stress-related parameters.

Figure 1. Wild-type [agouti] male Mongolian gerbil (*Meriones unguiculatus*).

Gerbil Morphology and its Implication for Species-adequate Husbandry

The Mongolian gerbil (*Meriones unguiculatus*) is a non-murine rodent of intermediate size between a rat and a mouse [body length 10-12.5 cm, adult weight 70-130 g]. The wild type has a golden brown, agouti-speckled coat on back and face and a cream-colored belly with a clear demarcation line between the two (Figure 1). Males tend to have a reddish tinge in the fur and are generally more heavily built than females. The gerbil has a fur-covered tail almost as long as the body [9.5-11 cm] with a small tuft of black hair at the tip. Since the first coat color mutations, "black" and "spotted," a wide variety of coat-colors has emerged. Differences in behavior and physiology between the different color varieties are not pronounced (Dizinno and Clancy, 1978). The hind legs of gerbils are relatively long compared to the forelegs due to longer hind paws. The animals often sit on their haunches to look around, feed or groom themselves, and the long hind paws are used extensively during digging behavior to vigorously kick out and remove loose substrate from under the belly. Gerbils perform rapid foot-thumping with the hind feet during alarming situations, and the males foot-thump too after copulation (Holman and Seale 1991). In erect, attentive posture, gerbils stand up on the toes, and the tail acts as extra support. An erect adult gerbil is about 12 cm high. Gerbils are not well equipped to climb, at least not upside down at the cage top as mice do (Roper and Polioudakis, 1977). This is due to the fact that their hind paws are not equipped with friction pads and partially opposable toes, and that the soles are covered with fur. Gerbil cages, however, have to be covered because the animals might crawl or jump out. If startled, gerbils can readily jump over a 30 cm high barrier. Adult animals can comfortably crawl through tubes of a diameter of 5 cm. Tubes narrower than 4 cm, however, may pose a problem for the more massively built adult males who might get stuck.

Gerbil claws grow too long if not worn down by regular digging, or walking on rough surfaces. Gerbil **teeth** also over-

Figure 2. Fur-clipping in an agouti gerbil—a method to mark animals individually.

grow quickly if not worn down by regular gnawing on hard food or chewing material. The animals are highly motivated to shred gnawable material such as cardboard, hay, straw, woodsticks (Glickman et al., 1967), and they should always have access to it. They shred such material into small pieces and afterwards collect it and carry it to the nest site to build or pad the nest. Wood chip bedding is too finely grained to trigger gnawing behavior.

Gerbil **tails** are well covered with fur. Gerbils should not be picked up by the tail, because they can shed the outer layer of skin and hair of the tail if in danger and escape with the bloody tail remains of bones and muscle. Dangling a gerbil from his/her tail base is an inhumane practice that can induce an epileptic-like seizure (Tumblebrook Farm, 1980; Fenske, 1996). The **marking** of same-colored individuals can be done by cutting away the brown upper, light fur layers in defined patterns. The dark under-fur will then show and allow for an easy recognition of individuals (Figure 2). This simple marking method is quite durable in adult gerbils (4-

8 weeks). Transpondering might offer a more sophisticated identification method, which has been used with success in house mice and in much smaller species such as insectivorous bats (Neuhäusser-Wespy and König 2000).

To allow for normal locomotion, the **flooring** of the laboratory cage must be covered with substrate or at least be somewhat rough, not just wire or slippery polycarbonate. Most small rodents, including gerbils, prefer bedding to wire or bare plastic as cage floor (Pettijohn and Barkes, 1978; Arnold and Estep, 1994; Patterson-Kane et al., 2001). The *Guide for the Care and Use of Laboraory Animals,* therefore, recommends "solid-bottom caging, with bedding... for rodents" (National Research Council, 1996, p. 24). We tested several **bedding** types and found that the gerbils' fur became greasy and lost its shine when the animals were kept on sawdust or dry sand. Gerbils kept on rough-grained sawdust, woodchips, hay, straw, or moist sand had shiny, healthy-looking fur. This was particularly noticeable in animals who had access to moist sand in which they could dig tunnels (own unpublished observation). As an alternative to the heavy moist sand, rough-grained, well-absorbing wood chips provide a suitable bedding for gerbils. The bedding should have a depth of 3-5 cm to allow for digging, and it should always be supplemented with hay or straw to provide more structure and opportunities for gnawing. The addition of a rough-surfaced object [e.g., a stone or the backside of a bathroom-tile] helps wear down the claws and provides a suitable object for marking behavior (Roper and Polioudakis, 1977; Thiessen and Yahr, 1977; Dizinno and Clancy, 1978). Gerbils should be able to stand up in **fully erect posture**. Therefore, the top of the cage should be slightly higher than an erect gerbil [12 cm] plus the depth of the bedding layer [3-5 cm]. This means that **gerbil cages must have a height of at least 17 cm.**

Gerbils have a relatively broad snout (cf., Figure 7). They have problems feeding on pellets presented in the food hopper with standardized 0.7 cm bar spacing, which is suitable for rats and mice. Competition between cage-mates increases when food is not so easily obtainable due to narrow bar spacing. Subordinates who manage to retrieve a food item from the food hopper are often chased and pestered by dominant cage mates (Waiblinger, unpublished data). Cage-tops with **0.9—1.2 cm bar spacing** are preferable for gerbils. Unfortunately such cage-tops are not commercially available. However, standardized cage-tops can easily be altered for gerbils by removing every second bar of the food hopper. **Scattering the food on the bedding** of the cage circumvents the problems associated with the food hopper and also provides behavioral enrichment: the gerbils can now dig and search for food items in the substrate (Cheal, 1987; Forkman, 1996) as well as carry food around and hoard it (Agren et al. 1989b). We found that adult gerbils dedicate up to 30% of their active time to foraging for food that is scattered on the bedding of their cage. Thanks to the dry feces and concentrated urine of gerbils, the bedding doesn't become wet so quickly, and the scattered food will remain relatively clean.

Gerbil Ecology and its Implication for Species-adequate Husbandry

The Mongolian gerbil can be found in all types of semi-desert in Mongolia, Inner Mongolia [China], Manchuria [China] and South Russia. Gerbils are found in habitats as diverse as dry sandy steppes, moist river banks and grassy semi-desert salt pans, but also man-made artificial earth structures, as long as the ground is sandy and easy to dig out (Milne-Edwards, 1867; Bannikov, 1954). The climate in these regions is continental. A range of physiological and behavioral adaptations enable gerbils to survive under often extreme climatic conditions. Gerbils excrete highly concentrated urine and almost dry fecal pellets. Although wild gerbils do not need to take up large amounts of water, they might lick dew or otherwise get their fluid through plant matter (Bannikov, 1954). In the wild, gerbils feed on the green parts of wormwood (*Artemisia* spp.), the seeds of numerous grasses, bulbs, buckwheat (*Fagopyrum*) and a wide range of other annuals and perennials (Bannikov, 1954; Naumov and Lobachev, 1975; Agren et al., 1989a). In the laboratory, gerbils cannot survive without water when they are fed only dry food (Marston and Chang, 1965; Tumblebrook Farm, 1975). The older a gerbil gets, the more water he/she needs (Spangler et al., 1997). Gerbils will drink water even if they are regularly provided with fresh fruit and vegetables. It is therefore advisable to **always provide them with fresh water**. Fruit [pears, melons with seeds, apples, oranges], vegetables [cucumber, carrot, pumpkin, fennel] and salads are readily accepted and enrich the gerbils' menu. Experience shows that gerbils prefer sweet and/or juicy produce. Since gerbils tend to hoard everything that is edible, cages need to be checked daily to remove any rotten fruit or vegetable material.

Although gerbils do not have cheek pouches like hamsters, they exhibit a very pronounced **hoarding behavior**. In the wild, 2-3 oval food stores per burrow have been found of 30-75 cm length and 15-20 cm height, containing green plant matter and seeds (Bannikov, 1954). Gerbils may hoard up to 1.5 kg of grains in one store (Bannikov, 1954; Naumov and Lobachev, 1975). In the laboratory, even under controlled temperature, humidity and day-length conditions, gerbils engage in extensive hoarding behavior (Tumblebrook Farm, 1977; Agren et al., 1989b; Forkman, 1996).

In their natural habitat, gerbils dig out vast subterranean

Figure 3. Gerbils in a seminatural burrow.

Figure 4. Gerbil engaged in stereotypic digging.

burrows up to 170 cm deep into the ground. These labyrinths may extend horizontally over 6 to 8 meters. They consist of a complicated net of tunnels and nest chambers, food stores and litter nests (Figure 3; Bannikov, 1954; Naumov and Lobachev, 1975; Roper and Polioudakis, 1977; Agren et al., 1989a). The burrow serves as retreat and refuge from predators, food-storage and protection from extreme temperature and desiccation. If a gerbil perceives a threat, he or she will thump the hind feet in a fast staccato, and this will prompt the whole group to rush to the various entrances of the borrow and disappear. Gerbils show surface activity day and night throughout the summer. During the winter they hide in the burrow and emerge only on the most sunny days (Bannikov, 1954; Naumov and Lobachev, 1975; Agren et al., 1989a,b). In summer, the burrows probably provide more even temperatures and air humidity, whereas they isolate against the cold in winter (Bannikov, 1954; Naumov and Lobachev, 1975).

Gerbils use a wide range of shredded, dry plant material to isolate and pad their nest chambers (Bannikov, 1954). They exhibit **shredding behavior** when presented with hay, straw, paper, cardboard, branches or wood sticks, collecting the shredded pieces and using them to build a nest (Glickman et al., 1967; Waiblinger and König, 1999). Most of these materials can be autoclaved, are cheap and provide the animals with excellent enrichment material. Gerbils are proficient shredders who reduce a cardboard tube to dime-sized pieces within 5 minutes! It is therefore important to regularly replenish the gnawing material with new supply to keep the animals busy. Paper and cardboard should not contain ink. Empty toilet paper rolls, egg containers or plain cardboard boxes are ideal shredding materials. Material from willow (*Salix* spp.), hazel (*Corylus avellana*), beech (*Fagus sylvatica*), birch (*Betula pendula*), maple (*Acer* spp.), elder (*Sambucus* spp.), fruit trees, as well as pine trees can provide natural gnawing sticks that are very attractive for gerbils. In the case of pine tree material, an animal will skillfully gnaw off all the needles before tackling and shredding the branch itself. It goes without saying that all poisonous plants such as ivy (*Hedera helix*), yew (*Taxus baccata*) and holly (*Ilex* spp.) have no place in the gerbil cage.

Under laboratory housing conditions, gerbils develop **stereotypic digging** behavior (Figure 4). Neither additional space allowance [7600 cm^2 instead of 1900 cm^2 for a family of gerbils] (Wiedenmayer, 1996) nor the performance of digging in natural substrate [dry, heavy sand] (Wiedenmayer, 1997a) is effective in preventing the development of stereotypic digging behavior in juveniles. However, the presence of a burrow during ontogeny results in significantly reduced stereotypic digging (Wiedenmayer, 1995,1997b).

The ordinary cage set-up does not allow for digging a stable, dark burrow system. Nevertheless, the gerbils try to dig a burrow in the lab cage, but this behavior strategy is not successful. Yet, the digging continues with great perseverance and gradually develops into a persisting stereotypy, indicating that the animals' behavior control mechanism is chronically overtaxed (Wiedenmayer, 1997a,b). **Ideally gerbils should be able to construct their own burrow.** In standard laboratory cages, this is impossible because they are not high enough to contain a sufficiently thick layer of bedding in which a stable burrow can be constructed by the animals. Also, most of the bedding will be kicked and shoved out of the cage, since the walls are not high enough.

Spacious, 50 cm high terrariums with wire-mesh tops are the alternative. Wood-chip bedding, mixed with hay, straw, cardboard boxes/tubes and branches provides a stable substrate. It needs to be filled in 20-30 cm high to allow the gerbils to dig their own complex burrow. Such terrariums are very suitable for breeding pairs or same-sex groups. Even with the opportunity to retreat into their burrow, most gerbils will nevertheless inquisitively emerge from it in the presence of familiar humans. Individuals can be swiftly caught and transported in a cardboard tube, as they have an innate urge to dive into the nearest tube when they feel threatened by personnel.

If it is necessary to have regular access to the animals, **artificial burrows** can be used to house gerbils. Wiedenmayer (1995) developed a simple artificial burrow system consisting of an opaque nestbox placed <u>outside</u> the cage. The animals have access to it through a 20 cm long tube inserted through the back wall of the cage. This *quasi* external burrow—unlike a dark nestbox attached to the cage without interconnecting tube—reduces stereotypic digging significantly (Wiedenmayer, 1995,1997a,b). For practical reasons, an artificial burrow system should be integrated into a standard laboratory cage rather than attached to it. For a laboratory cage type IV [38 x 58 cm x 20 cm height] the system consists of three modules (Figures 5 and 6a,b):

- An 18.5 cm high transparent separation wall, which divides the cage into the burrow area and an activity area with food hopper and water bottle. Attached to the separation wall is a transparent chamber [12.5 x 12.5 cm]. It helps to prevent conflicts of use between nest, food store and latrine (Müller, 1998).
- An opaque access tunnel [5 cm in diameter; total length approximately 34 cm] with a right angle to minimize light leakage into the nestbox and to save space. The tunnel can be constructed from plastic, but metal is preferable, since this element is the only one susceptible to gerbil gnawing.
- An opaque nestbox [12.5 x 19 cm] with removable lid, which is held in place by the cage-top pressing down on it.

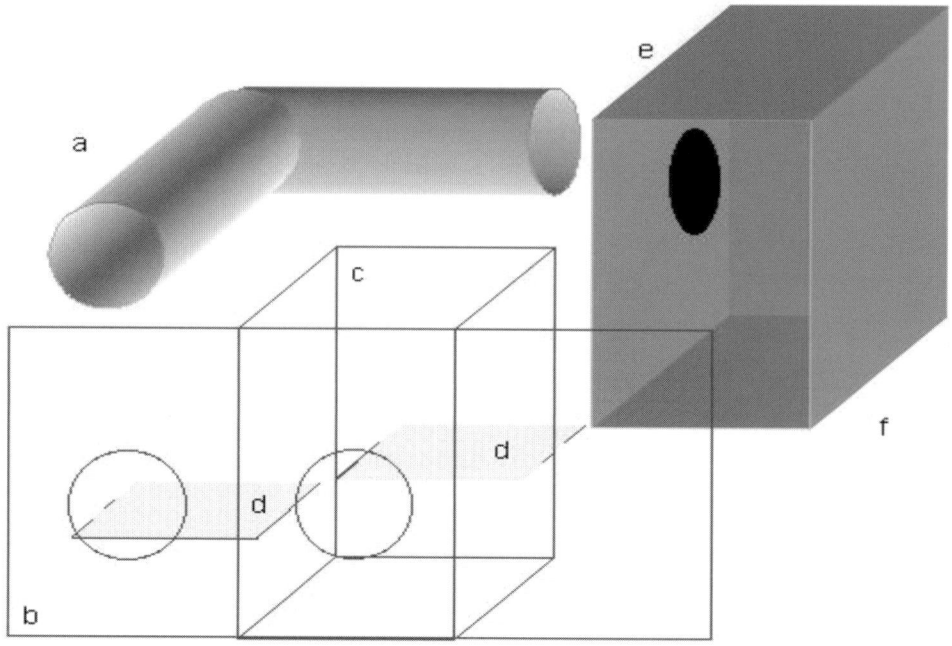

Figure 5. Modules of the artificial burrow system: opaque access tunnel (a), transparent separation wall (b) with transparent box (c) and supports (d) for access tunnel, opaque nestbox (f) [with front wall removed, and with lid (e)].

The three modules can be separately cleaned or replaced. Their integration in a cage type IV does not affect stacking density.

Gerbils who grow up in this opaque, artificial burrow system develop significantly less stereotypic digging than gerbils who grow up in a transparent burrow (Waiblinger and König 2001). A drawback of letting the animals grow up in an opaque burrow is their pronounced shyness resulting from the dark, secluded living environment. On the other hand, the burrow makes it possible to remove the animals from the cage and easily transport them within the nestbox without disturbing them unduly. The animals should not be taken out of the nestbox, since this experience is likely to make them avoid the burrow system altogether thereafter! Lifting the lid of the nestbox to check on the animals without handling them has proved to be not that problematic. Patient training with positive reinforcement helps taming and training the animals for routine handling. The following daily procedure assures that gerbils can readily be picked up:

- By sliding the top 5-10 cm backwards, the cage opens just enough for a gerbil to stand up and nose about.
- An animal emerges from the burrow, comes forward and inquisitively peeks out and sniffs (Figure 7).
- The animal is tamed by offering preferred food.
- Once tame, the animal is gently lifted out of the cage, returned back to the cage and rewarded with sunflower seeds. This exercise is repeated on different occasions until the animal promptly comes forward and accepts being picked up and held in the hand for a short while.

Gerbil Social Organization and its Implication for Species-adequate Husbandry

Gerbils are social animals. In the wild, group sizes range from 2 to 15 animals of all ages and sexes. Groups are founded by a breeding pair and extended by their offspring and other relatives (Bannikov, 1954; Naumov and Lobachev, 1975; Agren et al., 1989a). There is usually only one reproductively active female per group. Other adult, subordinate females are sexually suppressed (Norris and Adams, 1974; Payman and Swanson, 1980; Salo and French, 1989). The reproductive, dominant male is as involved in territorial defense and pup-care as the female, with the exception of retrieving pups (Waring and Perper, 1980; Weinandy and Gattermann, 1999). Older, already weaned offspring take the role of "alloparents" or "helpers" who participate in the care of younger siblings. By doing so they learn parental skills that will make them proficient parents. Breeding pairs with at least one alloparenting-experienced partner successfully raise more offspring than inexperienced pairs (Salo and French, 1989). Group members join forces in defending and marking the territory, constantly digging and extending the burrow, and hoarding food and nesting material (Agren et al., 1989a,b).

The highly social gerbils should never be kept alone! **Group-housing** is a simple way of effective environmental enhancement and welfare improvement. Breeding pairs can be kept together uninterruptedly without risk for their whole reproductive lifetime (Agren and Meyerson, 1977; Gerling and Yahr, 1978). The offspring should remain with the parents as long as there is enough space. They will not reproduce or show aggression under this housing condition (Ostermeyer and Elwood, 1984; Salo and French, 1989). A family group with two generations of offspring should never be kept in a [small] cage but in a large terrarium. Besides keeping gerbils in breeding pairs or family groups, same-sex groups are also possible. It doesn't matter whether males or females are kept together. Both social arrangements work well. It should be emphasized that adult gerbil males are not aggressive with each other, provided they know each other since they were juveniles.

The **formation of a new group** can be problematic. Our own experience shows that adult animals, especially females, are not likely to accept any strange partner(s) of either sex, except juveniles less than five weeks of age. We had best results with the formation of isosexual groups of eight-week-old animals who were introduced in cages with fresh bedding [neutral territory] and tubes for escape. We also noticed that powdering of the animals with a non-toxic, dry animal shampoo can help camouflage previous group odors thereby buffering territorial antagonism among the members of a newly established group. Careful observation of the animals during the first day after group formation helps to remove individuals who are persistent targets of overt aggression. Affiliative behaviors such as nose-to-nose contact, mutual grooming, sniffing each others' corners of the mouth, and nudging the head beneath each others' rump are signs that the new group is compatible.

Separating juvenile gerbils from their parents can also pose a problem. Despite our initial hypothesis that stereotypical **bar-chewing** may result from a lack of chewable nesting material or from reinforced bar-manipulation, caused by the close proximity of food pellets and cage bars in the food hopper, our experiments showed that neither of these factors has an influence on the development of this behavior pattern. Bar-chewing, however, significantly increases when juveniles are separated from their families <u>before</u> the next litter is born and housed in separate cages with <u>fresh bedding material</u> (Waiblinger and König, 1999; Waiblinger and König 2001). This suggests that juvenile gerbils should not be separated from their parents before a new litter is born and that at least part of the bedding in the new cage should originate from the parental cage.

Gerbil Scent Marking and its Implication for Species-adequate Husbandry

Gerbils of both sexes have a strong urge to scent mark their territory (Roper and Polioudakis, 1977; Yahr, 1977; Agren et al., 1989b). They deposit sebum from the ventral scent gland on stable objects and surfaces by pressing their bellies against them and walking over them. Prominent objects and territory borders may also be marked by using fecal pellets and urine (Agren, 1976; Thiessen and Yahr, 1977). Gerbils recognize group members by their scent (Halpin, 1975). Females and males show strong xenophobic reactions towards unfamiliar animals or their scent marks by grinding their teeth and lashing their tails. If not separated in time, two strange gerbils will attack and possibly kill each other during the fight.

Cage cleaning is usually a very disturbing experience for gerbils, probably because the fresh bedding is void of any familiar scent marks. After cage cleaning, gerbils typically show acute stress reactions (Weinandy, 1995) and an enhanced susceptibility to diseases (Peckham et al., 1974; Bucklar, personal communication). They need a much longer

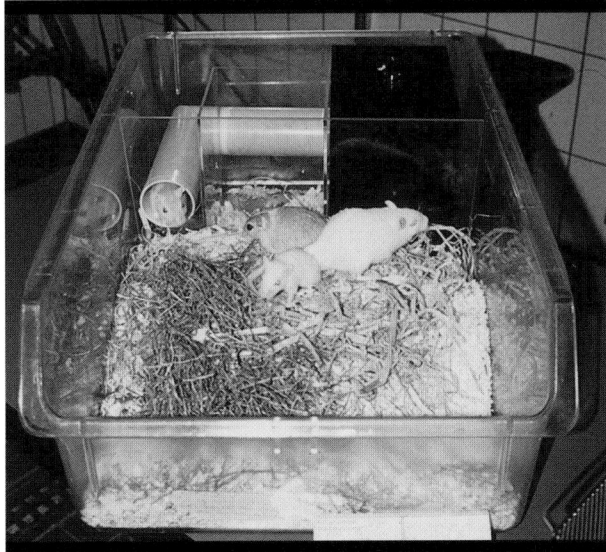

Figure 6a,b. Gerbil-inhabited laboratory cages type IV with integrated artificial burrow system, with top removed and top in place.

Figure 7. Gerbils are inquisitive and will emerge from a burrow [artificial or self-dug] in the presence of a familiar person. Using positive reinforcement, they can easily be tamed and trained to allow being picked up and held in a cupped hand.

heart-rate recovery time after cage cleaning than after handling or confrontation with an unfamiliar adult conspecific (Weinandy and Gattermann, 1995). Our own experience shows that these effects can be slightly buffered by leaving some soiled bedding [i.e., familiar odor] during the cage cleaning process.

Conclusions

To create adequate quarters for gerbils in research laboratories the following provisions must be made to address the animals' basic conditions for well-being:
- Species-adequate artificial burrow system;
- Social housing;
- Offsprings stay long enough in the family group to get experienced in the care of juveniles.

References

Agren G 1976. Social and territorial behavior in the Mongolian gerbil (*Meriones unguiculatus*) under seminatural conditions. Biology of Behavior 1, 267-285

Agren G, Meyerson BJ 1977. Influence of gonadal hormones on the behavior of pair-living Mongolian gerbils (*Meriones unguiculatus*) towards the cagemate versus a non-cagemate in a social choice test. Behavioral Processes 2, 325-335

Agren G, Zhou Q, Zhong W 1989a. Ecology and social behavior of Mongolian gerbils, *Meriones unguiculatus*, at Xilinhot, Inner Mongolia, China. Animal Behaviour 37, 11-27

Agren G, Zhou Q, Zhong W 1989b. Territoriality, cooperation and resource priority: hoarding in the Mongolian gerbil, *Meriones unguiculatus*. Animal Behaviour 37, 28-32

Arnold CE, Estep DQ 1994. Laboratory caging preferences in golden hamsters (*Mesocricetus auratus*). Laboratory Animals 28, 232-238

Bannikov AG 1954. The places inhabited and natural history of *Meriones unguiculatus*. Trudy Mongol'skoi Komissii 53, 410-415

Cheal M 1987. Environmental enrichment facilitates foraging behavior. Physiology & Behavior 39, 281-283

Dantzer R 1991. Stress, stereotypies and welfare. Behavioural Processes 25, 95-102

Dizinno G, Clancy AN 1978. Ventral marking in black and agouti gerbils (*Meriones unguiculatus*). Behavioral Biology 24, 545-548

Fenske M 1996. Dissociation of plasma and urinary steroid values after application of stressors, insulin, vasopressin, ACTH, or dexamethasone in the Mongolian gerbil. Experimental and Clinical Endocrinology and Diabetes 104, 441-446

Forkman B 1996. The foraging behavior of Mongolian gerbils: a behavioral need or a need to know? Behaviour 133, 129-143

Gerling S, Yahr P 1978. Effect of the male parent on pup survival in Mongolian gerbils (*Meriones unguiculatus*). Animal Behaviour 27, 310-311

Glickman SE, Fried L, Morrison BA 1967. Shredding of nesting material in the Mongolian gerbil. Perceptual and Motor Skills 24, 473-474

Halpin ZT 1975. The role of individual recognition by odours in the social interactions of the Mongolian gerbil (*Meriones unguiculatus*). Behaviour 58, 117-129

Holman SD, Seale WTC 1991. Ontogeny of sexually dimorphic ultrasonic vocalisations in Mongolian gerbils. Developmental Psychobiology 24, 103-115

Hughes BO, Duncan IJH 1988. The notion of ethological "need," models of motivation and welfare. Animal Behaviour 36, 1696-1707

Marston JH, Chang MC 1965. The breeding, management and reproductive physiology of the Mongolian gerbil (*Meriones unguiculatus*). Laboratory Animal Care 15(1), 34-48

Mason G 1991. Stereotypies and suffering. Behavioural Processes 25, 103-115

Milne-Edwards A 1867. Observations sur quelques mammifères du nord de la chine: *Gerbillus unguiculatus*. Annales des Sciences Naturelles (Zoologie) 5(7), 375-377

Müller L 1998. The influence of a transparent artificial burrow system on the ontogeny of digging behavior in Mongolian gerbils (*Meriones unguiculatus*). Masters-Thesis. University of Zürich, Zürich, Switzerland

National Research Council 1996. Guide for the Care and Use of Laboratory Animals, 7th Edition. National Academy Press, Washington, DC
Full Text: http://www.nap.edu/readingroom/books/labrats/

Naumov NP, Lobachev VS 1975. Ecology of the desert rodents of the USSR (Jerboas and Gerbils). In Rodents in Desert Environments Prakash I, Ghosh PK (eds), 529-536. Dr. W. Junk b.v. Publishers, The Hague, Netherlands

Neuhäusser-Wespy F, König B 2000. Living together, feeding apart: How to measure individual food consumption in social house mice. Behaviour Research Methods, Instruments & Computers 32, 169-172

Norris ML, Adams CE 1974. Sexual development in the Mongolian gerbil (Meriones unguiculatus), with particular reference to the ovary. Journal of Reproduction and Fertility 36, 245-248

Ödberg F 1989. Behavioral coping in chronic stress conditions. Behavioral and Social Sciences 48, 229-235

Ostermeyer MC, Elwood RW 1984. Helpers(?) at the nest in the Mongolian gerbil, Meriones unguiculatus. Behaviour 91, 61-77

Patterson-Kane WG, Harper N, Hunt M 2001. The cage preferences of laboratory rats. Laboratory Animals 35, 74-79

Payman BC, Swanson HH 1980. Social influence on sexual maturation and breeding in the female Mongolian gerbil (Meriones unguiculatus). Animal Behaviour 28, 528-535

Peckham JC, Cole JR, Chapman WL, Malone JB, McCall JW, Thompson PE 1974. Staphylococcal dermatitis in Mongolian gerbils (Meriones unguiculatus). Laboratory Animal Science 24, 43-47

Pettijohn TF, Barkes BM 1978. Surface choice and behavior in adult Mongolian gerbils. The Psychological Record 28, 299-303

Roper TJ, Polioudakis E 1977. The behavior of Mongolian gerbils in a semi-natural environment, with special reference to ventral marking, dominance and sociability. Behaviour 61, 205-237

Salo AA, French JA 1989. Early experience, reproductive success, and development of parental behavior in Mongolian gerbils. Animal Behaviour 38, 693-702

Spangler EL, Hengemihle J, Blank G, Speer DL, Brzozowski S, Patel N, Ingram DK 1997. An assessment of behavioral aging in the Mongolian gerbil. Experimental Gerontology 32, 707-717

Stuermer IW 1998. Reproduction and developmental differences in offspring of domesticated and wild Mongolian gerbils (Meriones unguiculatus). Zeitschrift für Säugetierkunde 63 (Sonderband), 57-58

Thiessen DD, Yahr P 1977. The Gerbil in Behavioral Investigations. University of Texas, Austin, TX

Tumblebrook Farm I 1975. Gerbil Care and Maintenance. The Gerbil Digest 2(2), 4

Tumblebrook Farm I 1977. The Mongolian gerbil in behavioral studies. The Gerbil Digest 4(3), 4

Tumblebrook Farm I 1980. The gerbil as a stroke model. The Gerbil Digest 7(2), 4

Waiblinger E, König B 1999. Do the presence of nesting material and the location of the food presentation have an effect on the development of bar-chewing in laboratory gerbils? Aktuelle Arbeiten zur artgemässen Nutztierhaltung, KTBL-Schrift 391, 178-186

Waiblinger E, König B 2001. Housing and husbandry affect stereotypic behavior in laboratory gerbils. 3R-Info-Bulletin 16
FT: http://www.forschung3r.ch/de/publications/bu16.html

Waring A, Perper T 1980. Parental behavior in Mongolian gerbils (Meriones unguiculatus) II Parental interactions. Animal Behaviour 28, 331-340

Wechsler B 1995. Coping and coping strategies, a behavioral view. Applied Animal Behaviour Science 43, 123-134

Weinandy R 1995. Untersuchungen zur Chronobiologie, Ethologie und zu Stressreaktionen der Mongolischen Wüstenrennmaus, Meriones unguiculatus. Doctoral-Thesis, Martin Luther-Universität, Halle-Wittenberg, Germany

Weinandy R, Gattermann R 1995. Measurement of physiological parameters and activity of a Mongolian gerbil during gravidity and lactation with an implanted transmitter. Physiology & Behavior 58, 811-814

Weinandy R, Gattermann R 1999. Parental care and time sharing in the Mongolian gerbil. Zeitschrift für Säugetierkunde 64, 169-175

Wiedenmayer C 1995. The ontogeny of stereotypies in gerbils. Doctoral-Thesis, Universität Zürich, Zürich, Switzerland

Wiedenmayer C 1996. Effects of cage size on the ontogeny of stereotyped behavior in gerbils. Applied Animal Behaviour Science 47, 225-233

Wiedenmayer C 1997a. Causation of the ontogenetic development of stereotypic digging in gerbils. Animal Behaviour 53, 461-470

Wiedenmayer C 1997b. Stereotypies resulting from a deviation in the ontogenetic development of gerbils. Behavioural Processes 39, 215-221

Winterfeld KT, Teucert-Noodt G, Dawirs, RR 1998. Social environment alters both ontogeny of dopamine innervation of the medial prefrontal cortex and maturation of working memory in gerbils (Meriones unguiculatus). Journal of Neuroscience Research 52, 201-209

Würbel H 2000. Behaviour and the standardization fallacy. Nature Genetics 26, 263

Würbel H 2001. Ideal homes? Housing effects on rodent brain and behavior. Trends in Neurosciences 24, 207-211

Yahr P 1977. Social subordination and scent marking in male Mongolian gerbils (Meriones unguiculatus). Animal Behaviour 25, 292-297

Eva Waiblinger is a zoologist with a strong interest in ethology and animal welfare. She wrote her dissertation on laboratory gerbil welfare at the University of Zürich, Switzerland. She is currently working for the Swiss Animal Protection (Schweizer Tierschutz) in the area of companion animal information and services.

Comfortable Quarters for Rats in Research Institutions

Monica M. Lawlor

Psychology Department, Royal Holloway, Egham, Surrey, TW2 OEX, United Kingdom

Where rats are used for scientific study it is of prime importance to avoid cruelty and to curtail the suffering that might be associated with experimental procedures. The humane care of animals, however, ought to go beyond that to an <u>active</u> attempt to promote comfort and well-being.

Although rats have lived with man for millennia, people in general know very little about them. Rats have an "image problem": very few people like them and even fewer love them. Generally rats are seen as disease carrying vermin to be exterminated, rather than creatures to be cherished. When one adds to this the consideration that rats are phenomenally adaptive, with a remarkable ability to withstand efforts to eliminate them, it is not surprising that they are capable of growing, living, and breeding in conditions that are far from ideal. They are great survivors.

Some excellent books cover the proper care of laboratory rats, but their emphasis is more often on experimental procedures and the physiological aspects of care than on the quality of life of the animals who involuntarily make such a huge contribution—in terms of millions of animals "used" and "sacrificed"—to the advancement of biomedical science. In this chapter the emphasis is reversed, and the focus shifted on behavioral considerations rather than physiological data.

The currently recommended caging parameters for rats must be re-thought and revised if they cannot be shown to satisfy basic behavioral, physiological, and exercise needs of the animals who live in such enclosures. In order to decide the best way to satisfy those needs, it is logical to start by measuring the rat rather than the cage. A description of the rat's biology will make it clear why gross body weight should not be the only measure used to determine caging standards.

Species-typical Characteristics

Rats are long-tailed, conspicuously inquisitive rodents who are biologically adapted to live in groups. The most commonly used laboratory rats are mutants of the Gray Norway rat *(Rattus norvegicus)*. Selective breeding over hundreds of years made the mutant forms "domesticated," lacking the timidness and ferocity of their wild cousins. Their natural longevity is 3-4 years. Two types of red-eyed albinos are the prevailing laboratory rats: the relatively small, fine boned, and elegant Wistar rat; and the larger, relatively coarse boned and rough coated Sprague-Dawley rat. Pigmented, black-eyed rats are less common. The most widely used types are the Wistar Hooded and Lister Hooded rats. Their vision is far better than that of albino rats and they are less distressed by bright light.

Adult rats usually weigh between 200 and 1000 grams and measure 40 to 50 cm from nose to tail. When standing on all four feet, rats require extra horizontal space for their long, normally straight held tails. To assume a natural quadrupedal position an adult rat needs 35-48 cm. The bipedal position is very often seen in rats. It serves as an orienting stance in which the weight is on the back feet and the spine extended upward. The base of the tail is used as a stabilizing tripod (Figure 1). The forepaws may be supported on a firm vertical surface allowing the animal to stretch right up until he/she is standing on tiptoe. The head is up and the ears are pricked. The head room required for an adult animal to make the bipedal orienting stance is up to 30 cm (Figure 2).

Rats are social animals who live in stable groups in which each animal has a well-defined relationship with each member of the group. When kept in single-cages, rats suffer from isolation stress, which, in turn:

Comfortable Quarters for Laboratory Animals Reinhardt V, Reinhardt A (eds), 26-32. Animal Welfare Institute, Washington, DC 20007

Figure 1. Lister hooded male rats "challenging" each other in bipedal position.

- affects their growth, behavior, physiological condition, and responses to a wide variety of drugs (Baer, 1971);
- impairs their behavioral health as reflected in a higher incidence of stereotypical behavior patterns (Baenninger, 1967; Hurst et al., 1997); and
- reduces their survival rate when compared with group-housed animals (Shaw and Gallagher, 1984).

The presence of another rat has a protective effect in fear-provoking and stressful situations (Latané and Glass, 1968; Latané, 1969; Taylor, 1981).

Adult rats often groom each other, and by doing so spread a group-specific scent on each other with saliva. They also spend a considerable part of their time grooming themselves. They do this in the squatting or standing position with tongue, teeth, paws, and claws. Ano-genital sniffing among same-sex animals is often a prelude to a dispute. At this point a fight may be averted if one animal rolls over onto his or her back and emits a submission call. If this happens, the dominant will stand over the other and do nothing more; both may even fall asleep in this position. If, however, neither animal gives way to the other, both will rear up and start to box with their forepaws while making attempts to bite each other. Any such encounter may be ended by a submissive posture and call. If this fails to happen, the animals may injure each other badly. Overt fighting can be a serious problem when rats are kept in barren cages, because the defeated animal cannot effectively stop the attacks of the aggressor by running away and escaping from the opponent's sensory field.

Parturient females have a strong need for nesting material and will readily work for the acquisition of it (Oley and Slotnick, 1970). Apart from females approaching parturition, rats do not show much purposive nest building activity. Preferred sleeping sites are "nest-like" but do not usually have much structure.

Mating is largely opportunistic. When in full oestrus the female becomes active and accepts copulation from any sexually competent male. During coitus, ultrasonic calls are emitted by females, and males purr during mating. These vocalizations seem to serve to deter other group members from interfering with the mating couple.

The newborn not only need the attentive care of their mothers but also a distinctive nest to keep them warm and safe during the short periods when the mother leaves them. The mother carefully covers the pups when she leaves the nest. Sometimes, however, pups may explore the environment outside of the nest. Females—and to a lesser degree also males—have a strong drive to retrieve such pups even if the young are already too old to relish the attention.

When rats are comfortable and relaxed they normally nurse in the "half moon position." In less benign circumstances they nurse in a "cover position," standing over the pups in a protective manner (Figure 3). When there is sufficient nest material, and the ambient temperature is adequate, the mother will leave the nest between feeds. In more demanding conditions, when the temperature is too low, and there is insufficient nest material, she will remain crouched over the pups to keep them warm. If the caging system allows her to do so, the mother will "escape" from the increasingly mobile and demanding offspring once they are two or three weeks old; however, she will allow some sporadic nursing until the pups are four or five weeks old.

Young rats play a lot. Play is a form of vigorous exercise that is essential for the well-being and normal social and sexual development of young rats. Adult animals play only very rarely with each other.

Rats are proficient diggers and build their own tunnels in which they usually sleep during the light phase of the day.

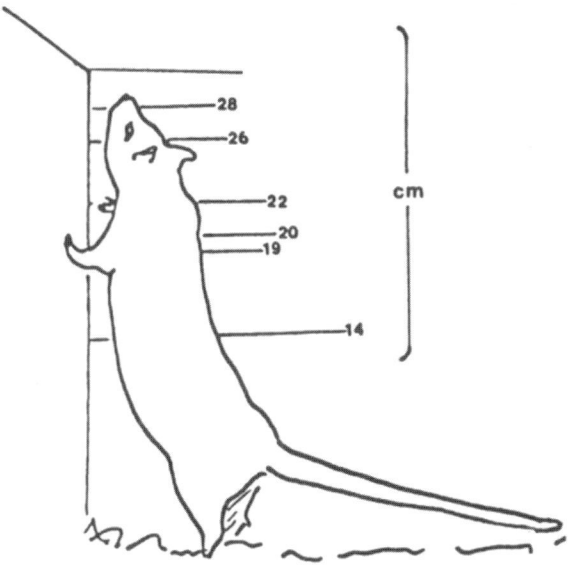

Figure 2. Adult rat in full bipedal posture, showing the headroom required.

Figure 3. Lister hooded rat nursing her ten-day-old pups in "cover position."

They will sleep in a heap or separately depending on the ambient temperature. They are easily aroused and do not normally sleep for long periods without waking intervals. Mature animals settle down to sleep by tucking their heads between the forepaws while in the quadrupedal position. As they move into a deeper sleep phase, they suddenly keel over onto a side, typically at full length with tail extended. To assume this natural sleeping posture an adult rat requires a floor space of 15 x 35-45 cm. Rats sleep in a curled position only when chilled. When waking up, they stretch and yawn with fully opened mouth while the head is thrown back and the forelegs extended. When a rat stretches, one forepaw goes forward of the head while a back foot is stretched out beyond the tail base, the tail itself being arched; then the feet are reversed. Finally, the animal shakes itself.

Rats will progress by creeping when they are nervous, insecure, or alarmed. In this mode of locomotion the animal flattens his/her belly against the floor and shoves the trunk forward by "paddling" with laterally extended limbs. When a rat is walking normally, the tail is carried off the ground straight out behind the trunk. Any other pattern is distorted and indicative that the animal is kept in a too small enclosure. The running pace is probably three or four times longer than the walking pace. When a rat is running, the tail is carried straight out behind the body with the tip of the tail upturned. Juveniles and females in estrus spend conspicuous amounts of time running around. Rats can achieve considerable speed, especially in a panic-stricken dash. When merely exploring a new place they lope along at a much more modest speed.

The motivation to forage is very strong in rats, and they will readily work for the retrieval of food in the presence of freely accessible identical food (Neuringer, 1969; Carder and Berkowitz, 1970; Hothersall et al., 1973). They will eat powdered or mushy food from a dish or from the floor, but their species-specific habit is to secure a piece of food in their teeth and carry it to a suitable spot where they adopt a squatting posture and transfer the food to the forepaws. Holding the food in their paws, they nibble gently at it; if they do not like the taste they drop it immediately. The opportunity to gnaw is an essential physiological and behavioral need for rats. If they are not given the chance to regularly gnaw, their front teeth overgrow and make it more or less impossible for them to eat at all or to engage in grooming.

Rats have a spontaneous fear of people and avoid being handled. Handling can be a powerful stressor for them (Brown and Martin, 1974; Kvetnansky et al., 1978; Berkey et al., 1990; Briese and Cabanac, 1991) introducing uncontrolled variability into research data (Shyu et al., 1987; Brockway et al., 1993; Claassen, 1994). Rats are also very sensitive to environmental disturbances. Even ordinary animal husbandry procedures such as moving a cage to a different area or moving animals to a clean cage can induce transient, but significant, physiological and behavioral changes that may confound experiments conducted shortly thereafter (Gärtner et al., 1980; York and Regan, 1982; Saibaba et al., 1996; Duke et al., 2001).

The natural defense of rats who experience threat is not to hang about but to run and hide, and if possible huddle with other conspecifics in a safe place. Being placed on an open surface is an especially threatening situation (cf., Latané, 1969).

Minimum Recommendations for Rat-adequate Housing and Handling Conditions

The ethogram provides a base from which the behavioral needs of a species can be derived and that allows one to make recommendations regarding the minimum space and caging conditions required by the animals to satisfy those needs and experience a state of behavioral and physical well-being.

The cage in which a mature rat can adopt species-typical postures and stances and can carry out essential activities has to measure between 35 x 25 x 18 [height] cm for the smallest females and 50 x 30 x 30 [height] cm for the largest males. Table 1 lists the minimum space requirements by sex and body weight. It must be emphasized that young animals require more space, relatively, for play activities. Therefore, they should not be allocated less space than is appropriate for the smallest females [35 x 25 x 18 cm].

Rats of any age should not be caged singly or in large groups. For adults the group should not be more than six animals, for juveniles not more than ten animals. Rats kept in larger groups tend to be too aggressive and are more prone to disease. Pair-housing is probably the optimal alternative both to single-housing and to group-housing (Heath, 1999). Separation from conspecifics is a distressing situation for rats leading to significant physiological alterations (Ehlers et al., 1993; Young et al., 1996; Lawson and Churchill, 2000). Individually caged animals are susceptible to stress (Hurst et al., 1997), which again jeopardizes the validity of research data collected from such animals (Pérez et al., 1997). Rats show more pronounced stress-like changes in behavior and cardiovascular function during common husbandry and experimental procedures when they are housed alone than when housed with another rat (Zammit et al., 2001). If an

BODY WEIGHT OF RAT [1G=0.035 OZ.]	MINIMUM FLOOR AREA CM² (IN.²)		MINIMUM HEIGHT CM (IN)
	FOR 1-3 RATS	FOR AN ADDITIONAL RAT	
male up to 150 g female up to 140 g	900 (140)	300 (47)	18 (7)
male 150-250 g female 140-170 g	1200 (186)	450 (70)	20 (8)
male 250-450 g female 170-310 g	1500 (233)	600 (93)	22 (9)
male 450-900 g female 310-615 g	1800 (279)	800 (124)	26 (10)
male over 900 g female over 615 g	1800 (279)	1000 (155)	30 (12)

Table 1. Minimum space recommendations for laboratory rats.

animal has to be single-caged for veterinary reasons provision must be made that he/she can keep visual and auditory contact with other rats to buffer the stress associated with isolation (cf., Latané and Glass, 1968).

The main food staple for laboratory rats is ordinarily a commercial high quality pelleted diet fed ad libitum. Hard pellets usually provide for sufficient gnawing. **Natural food items, however, such as carrots, grain/seeds, and/or pieces of soft wood, are more species-appropriate items for gnawing.** Rats should always have free access to them. Wooden gnawing blocks are attractive enrichment objects (Chmiel and Noonan, 1996) that not only reduce the incidence of stereotypic chewing of metal cage bars (Orok-Edem and Key, 1994) and make the animals less timid (Eskola and Kaliste-Korhonen, 1998) but are available with certificates of analysis, a particularly important aspect for toxicological studies (Robertson, 1999). Rats "want" to forage (cf., Neuringer, 1969), and they can easily be induced to "work" for their food by soldering metal plates over their food hoppers, so that only a small segment of the original area remains available. This method of "food restriction" is preferable to giving less food to avoid obesity. Rather than rapidly eating a reduced ration and feeling hungry for long periods, the animals work harder for their food, which enables them to burn more calories and eat throughout the day. This reduces the incidence of obesity and its associated disorders and also encourages more "natural" behavior patterns, both of which improve welfare (Wrightson and Dickson, 1999).

In order to provide rats a sense of security and options of breaking visual contact with each other during agonistic conflicts, it is recommended to add vertical barriers (cf., Anzaldo et al., 1994) and/or tubes—made of PVC or aspen wood (Mering, 2000)—in their cages. This offers the animals additional wall contact, tactile comfort, escape routes, and areas for exploration, thereby increasing cage complexity and the usable floor space of the cage. Evidence suggests that a more complex housing environment—in sharp contrast to the barren cage—buffers anxiety responses to potential stressors (Levine, 1985). **A well-designed cage provides a distinctive sheltered nest area away from the feeding location.** Rats with access to an appropriate shelter are more explorative and less timid than those in barren cages (Townsend, 1997). Nest-boxes of opaque or semi-opaque materials are particularly suitable shelters (Manser et al., 1998). Ideally, rats should always have access to one cage section that is covered with a black perspex screen serving as dark-and-sheltered sleeping and hiding area and another section serving as living area (Figures 4; cf., Wrightson and Dickson, 1999). The living area section should be covered with a wire lid for gymnastics.

It should go without saying that solid floors are much more appropriate for the feet of rats than wire floors, which impact the feet in a biologically abnormal manner (Grover-Johnson and Spencer, 1981) and may cause discomfort, pressure sores, and pain. They may also cause chilling even in a warm room. While rats housed on grid or mesh floors

Figure 4. The ideal double-cage arrangement: The right cage section is covered with a black perspex screen and serves as dark-and-sheltered sleeping/nesting area. The left cage section serves as living area, which is covered with a wire lid for gymnastics.

tend to pile up in heaps when resting, rats with access to solid flooring spread out on the bedding (Wrightson and Dickson, 1999). Under experimental conditions, rats are prepared to make considerable efforts to reach a solid floor when they wish to rest. Preference testing revealed that the animals chose to dwell on solid floors rather than grids, regardless of previous housing experience. Thus **there is ground for suggesting that laboratory rats be housed on solid rather then grid floors, because solid floor housing improves their welfare** (Manser et al., 1995; Manser et al., 1996; Stauffacher, 1996). As a warning however, it must be pointed out that on solid PVC (polyvinyl chloride) floors the claws of rats may "over-grow" because the surface is not abrasive. If this happens, the claws must be clipped or the animal will experience considerable discomfort. The widely used concept of housing rats on one type of cage flooring should be abandoned and replaced by a cage concept with different types of flooring—and bedding—to enable the animals to express a more complete behavioral repertoire (van de Weerd et al., 1996; Wrightson and Dickson, 1999).

It is impossible to fully satisfy the instinctive need for digging in caged rats. Dust-free woodchip (preferably irradiated) bedding is a good compromise solution that allows the animals not only to engage in quasi-digging maneuvers but also to forage, i.e., search for food particles. The *Guide for the Care and Use of Laboratory Animals* aptly recommends "solid-bottom caging, with [emphasis added] bedding...for rodents" (National Research Council, 1996, p. 24). **Woodchips**—unlike sawdust—**and corn-cob are the preferred bedding for rats** (Blom et al., 1995; Patterson-Kane et al., 2001) **and should be regarded as a basic, inexpensive means of environmental enrichment.** The bedding also absorbs urine and moisture from feces. Regularly changed bedding is the best guarantee of a hygienic cage environment.

Parturient females must have access to shredded paper or soft wood to build appropriate nests for the successful rearing of their offspring (Figure 5). The use of shredded paper, which incidentally makes a nest similar to that of the wild rat, allows the female with her young to burrow and insulate themselves from disturbing environmental factors, thus enhancing the mother's feeling of security. Access to shredded paper can drastically reduce infant mortality (Nolen and Alexander, 1966).

All rats of a given experiment/test have to be caged at the same distance from the light source to assure uniform illumination. Traditionally, however, rat cages are arranged in racks with several rows being stacked on top of each other. This allows for maximal usage of room space but bears serious implications for scientifically valid research. Since the cages are arranged at different levels and most of them are located in the shade area cast by upper rows, it is impossible to guarantee uniform illumination in all cages. Light intensities in stacked cages vary substantially (Bellhorn, 1980; Clough, 1982). This introduces an uncontrolled variable into research data and unnecessarily increases the number of animals needed to obtain statistically significant results. The different degrees of illumination may be one of the explanations for variations in experimental results (Lockard, 1962; Weihe et al., 1969) and the need for unduly high numbers of animals to obtain statistically significant research findings. Obviously, the multi-tier caging systems are not compatible with scientifically sound research methodology.

Rats can lose their shyness of people if a little time is spent handling them as juveniles. An effective technique is to put half a dozen three to four weeks old cage mates into a bucket—using cupped hands to make the transfer from the cage—and then putting your hand into the bucket and allowing the young to explore it thoroughly. By putting your hand over and under the animals, they get used to the contact and can, in a day or two, be easily lifted a few inches in one hand. The bucket prevents the juveniles from avoiding the hand, thereby conditioning them to accept the human hand as a neutral, non-threatening environment. It is only necessary to do this exercise a few times over five days to "create" a rat who will readily accept proper handling throughout his or her lifetime (Figure 6). This simple technique—which is equally effective for mice and hamsters—is relatively labor-intensive, but it is reliable and makes life a great deal easier for both the rat and the handler. Gentle handling during infancy makes rats less fearful and quasi-tame in situations in which control rats remain timidly crouched at the back of the cage (Wells, 1985).

In an ideal world **all rodents are best handled by being picked up with a firm-and-gentle hold over the shoulders and quickly supported by allowing their feet to rest on your other hand or sleeve.** To a considerable extent proper handling depends on the handler rather than on the animal subject. Nervous people make animals [and other people] nervous and consequently unpredictable in their reactions to handling. Inexperienced personnel often grip too hard, thereby stressing not only the handled animal but also other rats who witness this disturbing situation (Fuchs et al., 1987). It is unusual for laboratory rats to object to being picked up in shoulder hold. Small rats fit comfortably into the hand when lifted and may be held in one hand if the tail is anchored between the third and fourth finger, and the thumb kept under the jaw. Bigger rats need the support of the second hand or a sleeve to make them feel secure when lifted.

Figure 5. A three-day-old litter of Lister hooded rat pups. The well-structured nest is built with shredded dye-free paper.

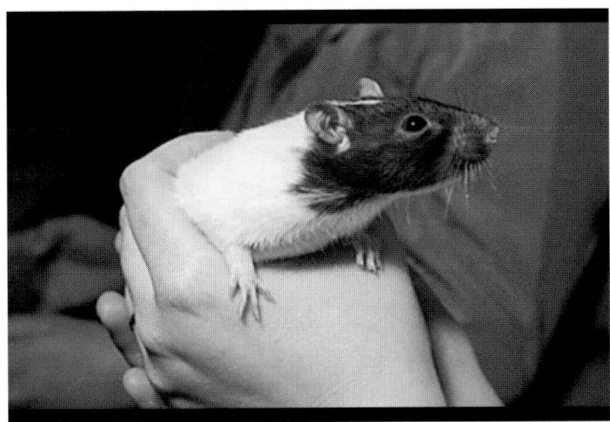

Figure 6. Once a rat is conditioned as a juvenile to accept the human hand as a neutral, non-threatening environment, he or she will readily accept proper handling throughout his or her lifetime.

Conclusion

Even though U. S. federal regulations currently do not regard rats as "animals" (United States Department of Agriculture, 1995), rats require and deserve the same professional care as other, perhaps more charismatic laboratory animals, because their well-being determines no less the quality and reliability of scientific research data collected from them. In order to design a species-appropriate and scientifically sound housing protocol it is essential to view the world with the eyes of a rat. It is through patient observation and a grain of humbleness that such a view can be cultivated.

References

Anzaldo AJ, Harrison PC, Riskowski GL, Sebek LA, Maghirang R, Stricklin WRGHW 1994. Increasing welfare of laboratory rats with the help of spatially enhanced cages. Animal Welfare Information Center (AWIC) Newsletter 5(3), 1-2 & 5
  Full Text: http://www.nal.usda.gov/awic/newsletters/v5n3/5n3anzal.htm

Baenninger LP 1967. Comparison of behavioural development in socially isolated and grouped rats. Animal Behaviour 15, 312-323

Baer H 1971. Long-term isolation stress and its effects on drug response in rodents. Laboratory Animal Science 21, 341-349

Bellhorn RW 1980. Lighting in the animal environment. Laboratory Animal Science 30, 440-450

Berkey DL, Meeuwsen KW, Barney CC 1990. Measurements of core temperature in spontaneously hypertensive rats by radiotelemetry. American Journal of Physiology 258, R743-749

Blom HJM, van Tintelen G, Bauman V, van den Broeck J, Beynen AC 1995. Development and application of a preference test system to evaluate housing conditions for laboratory rats. Applied Animal Behaviour Science 44, 279-290

Briese E, Cabanac M 1991. Stress hyperthermia: Physiological arguments that it is fever. Physiology and Behavior 49, 1153-1157

Brockway BP, Hassler CR, Hicks N 1993. Minimizing stress during physiological monitoring. In Refinement and Reduction in Animal Testing Niemi SM, Willson JE (eds), 56-69. Scientists Center for Animal Welfare, Greenbelt, MD

Brown GM, Martin JB 1974. Corticosterone, prolactin, and growth hormone responses to handling and new environment in the rat. Psychosomatic Medicine 36, 241-247

Carder B, Berkowitz K 1970. Rats' preference for earned in comparison with free food. Science 167, 1273-1274.

Chmiel DJ, Noonan M 1996. Preference of laboratory rats for potentially enriching stimulus objects. Laboratory Animals 30, 97-101

Claassen V 1994. Neglected Factors in Pharmacology and Neuroscience Research. Elsevier, Amsterdam, Netherlands

Clough G 1982. Environmental effects on animals used in biomedical research. Biological Reviews 57, 487-523

Duke JL, Zammit TG, Lawson DM 2001. The effects of routine cage-changing on cardiovascular and behavioral parameters in male Sprague-Dawley rats. Contemporary Topics in Laboratory Animal Science 40(1), 17-20

Ehlers CL, Kaneko WM, Owens MJ, Nemeroff CB 1993. Effects of gender and social isolation on electroencephalogram and neuroendocrine parameters in rats. Biological Psychiatry 33, 358-366

Eskola S, Kaliste-Korhonen E 1998. Effects of cage type and gnawing blocks on weight gain, organ weights and open-field behaviour in Wistar rats. Scandinavian Journal of Laboratory Animal Science 25, 180-193

Fuchs E, Fluegge G, Hutzelmeyer HD 1987. Response of rats to the presence of stressed conspecifics as a function of day time. Hormones and Behavior 21, 245-252

Gärtner K, Büttner D, Döhler R, Friedel J, Lindema J, Trautschold I 1980. Stress response of rats to handling and experimental procedures. Laboratory Animals 14, 267-274

Grover-Johnson N, Spencer PS 1981. Peripheral nerve abnormalities in aging rats. Journal of Neuropathology and Experimental Neurology 40, 155-165

Heath M 1999. Preliminary behaviour data for single and pair housed rats. Animal Technology 50, 47-48

Hothersall D, Huey D, Thatcher K 1973. The preference of rats for free or response-produced food. Animal Learning and Behaviour, 241-243

Hurst JL, Barnard CJ, Nevison CM, West CD 1997. Housing and welfare in laboratory rats: Welfare implications of isolation and social contact among caged males. Animal Welfare 6, 327-347

Kvetnansky R, Sun CL, Lake CR, Thoa N, Torda T, Kopin IJ 1978. Effect of handling and forced immobilization on rat plasma levels of epinephrine, norepinephrine, and dopamine-beta-hydroxylase. Endocrinology 103, 1868-1874

Latané B 1969. Gregariousness and fear in laboratory rats. Journal of Experimental Social Psychology 5, 61-69

Latané B, Glass D 1968. Social and nonsocial attraction in rats. Journal of Personality and Social Psychology 9, 142-146

Lawson DM, Churchill PC 2000. The effects of enrichment on parameters in hypertensive rats. Contemporary Topics in Laboratory Animal Science 39(1), 9-13

Levine S 1985. A definition of stress? In Animal Stress Moberg GP (ed), 51-69. Waverly Press, Baltimore, MD

Lockard RB 1962. Some effects of maintenance luminance and strain differences upon self-exposure to light by rats. Journal of Comparative and Physiological Psychology 55, 1118-1123

Manser CE, Broom DM, Overend P, Morris TH 1998. Investigation into the preference of laboratory rats for nest-boxes and nesting materials. Laboratory Animals 32, 23-35

Manser CE, Elliott H, Morris TH, Broom DM 1996. The use of a novel operant test to determine the strength of preference for flooring in laboratory rats. Laboratory Animals 30, 1-6

Manser CE, Morris TH, Broom DM 1995. An investigation into the effects of solid or grid cage flooring on the welfare of laboratory rats. Laboratory Animals 29, 353-363

Mering S 2000. Housing environment and enrichment for laboratory rats—refinement and reduction outcome. Natural and Environmental Sciences—Kuopio University Publications C 114, 1-60

National Research Council 1996. Guide for the Care and Use of Laboratory Animals, 7th Edition. National Academy Press, Washington, DC
Full Text: http://www.nap.edu/readingroom/books/labrats/

Neuringer AJ 1969. Animals respond for food in the presence of free food. Science 166, 399-401

Nolen GA, Alexander JC 1966. Effects of diet and type of nesting material on the reproduction and lactation of the rat. Laboratory Animal Care [Laboratory Animal Science] 16, 327-336

Oley NN, Slotnick BM 1970. Nesting material as a reinforcer for operant behavior in the rat. Psychonomic Science 21, 41-43

Orok-Edem E, Key D 1994. Responses of rats *(Rattus norvegicus)* to enrichment objects. Animal Technology 45, 25-30

Patterson-Kane EG, van de Ven M, Ras T 2001. Enrichment of laboratory rat caging. 2001 AALAS Official Program, 106

Pérez C, Canal JR, Dominguez E, Campillo JE, Guillén M 1997. Individual housing influences certain biochemical parameters in the rat. Laboratory Animals 31, 357-361

Robertson D 1999. Environmental stimulation for rodents on toxicological studies. Animal Technology 50, 182-183

Saibaba P, Sales GD, Stodulski G, Hau J 1996. Behaviour of rats in their home cages: daytime variations and effects of routine husbandry procedures analysed by time sampling techniques. Laboratory Animals 30, 13-31

Shaw DC, Gallagher RH 1984. Group or singly housed rats? In Standards in Laboratory Animal Management. The Universities Federation for Animal Welfare 65-70. The Universities Federation for Animal Welfare, Potters Bar, UK

Shyu WC, Nightingale CH, Tsuji A, Quintiliani R 1987. Effect of stress on pharmocokinetics of amikacin and ticarcillin. Journal of Pharmaceutical Sciences 76, 265-266

Stauffacher M 1996. Comparative studies on housing conditions. In Harmonization of Laboratory Animal Husbandry. O'Donoghue PN (ed), 5-9. Royal Society of Medicine Press, London, UK

Taylor GT 1981. Fear and affiliation in domesticated male rats. Journal of Comparative and Physiological Psychology 95, 685-693

Townsend P 1997. Use of in-cage shelters by laboratory rats. Animal Welfare 6, 95-103

United States Department of Agriculture 1995. Definition of terms under the Animal Welfare Act as amended (7 USC, 2131-2156). 9 CFR Ch. 1 (1-1-95 Edition), §1.1, 1-6
Full Text: http://www.access.gpo.gov/nara/cfr/waisidx_00/9cfr1_00.html

van de Weerd HA, van den Broek FAR, Baumans V 1996. Preference for different types of flooring in two rat strains. Applied Animal Behaviour Science 46, 251-261

Weihe WH, Schidlow J, Strittmatter J 1969. The effect of light intensity on the breeding and development of rats and golden hamsters. International Journal of Biometeorology 13, 69-79

Wells PA 1985. The Influence of Early Handling on the Temporal Sequence of Activity and Exploratory Behaviour in the Rat. University of London (PhD thesis), London, UK

Wrightson D, Dickson C 1999. Diet restriction through hopper design. Animal Technology 50, 45-46

York JL, Regan SG 1982. Conditioned and unconditioned influences on body temperature and ethanol hypothermia in laboratory rats. Pharmacology, Biochemistry and Behavior 17, 119-124

Young LA, Pavlovska-Teglia G, Stodulski G, Hau J 1996. Effect of group housing and oral corticosterone administration on weight gain and locomotor development in neonatal rats. Animal Welfare 5, 167-176

Zammit TG, Duke JL, Lawson DM 2001. Behavioral and cardiovascular responses of male Sprague-Dawley rats to common husbandry and experimental procedures: Effect of housing density. 2001 AALAS Official Program, 79

Monica Lawlor is a psychologist specialized in ethology. She maintained a rat colony for thirty years and conducted ground-breaking research in species-adequate housing arrangements and humane handling practices of rats kept in research laboratories. Monica Lawlor lives in London, United Kingdom, where she retired from the Psychology Department at Royal Holloway & Bedford New College.

Comfortable Quarters for Hamsters in Research Institutions

Gernot Kuhnen

Büro für Thermophysiologie, Beratung & Gutachten, Stettiner Straße 8, D-35415 Pohlheim, Germany; email: gernot.kuhnen@t-online.de

There are a number of species of hamsters, but the Syrian or golden hamster (*Mesocricetus auratus*) is the species that is most commonly found in research institutions. The entire laboratory and pet population of the golden hamsters appear to be the descendants of a single brother-sister pairing. These littermates were captured and imported from Aleppo [Syria] by Aharoni, a zoologist of the University of Jerusalem in 1930. In Jerusalem the hamsters were bred very successfully. Years later, animals of this original breeding colony were exported to the USA, where golden hamsters started to become one of the most popular pets and laboratory animals (Figure 1). Comparative studies of domestic and wild golden hamsters have shown that there is a drastically reduced genetic variability in the domestic strain. However, the differences in behavioral, chronobiological, morphometrical, haematological and biochemical parameters are relatively small and fall into the expected range of interstrain variations in other laboratory animals (Gattermann, 2000).

There are excellent books on the care of laboratory rodents, but they are dealing more with the demands of the experimenter than with the needs of the experimental animals. In this chapter, the focus is shifted and recommendations made that address the golden hamsters' basic behavioral and physiological needs under the constraints of confinement. To meet these needs is to assure the animals' well-being and the validity of research data collected from them.

Figure 1. In the United States more than 150,000 hamsters are used in research (U.S. Department of Agriculture, 2000; photo by E.P. Walker).

Species-typical Characteristics of the Golden Hamster

The original habitat of the golden hamster is the elevated plain around Aleppo in the northeast of Syria. The climate in the summer is hot and dry during the day and cold during the end of the night and the early morning. The winter is wet and cool, and there may be some days with snow and temperatures below 0°C.

Hamsters are solitary animals. They dig burrows, create nests and will hibernate if the temperature becomes low. Golden hamsters are "real" hibernators who lower their body temperature close to the ambient temperature [but not below 0°C]. This kind of thermoregulation drastically diminishes the metabolic rate to about five percent and helps the animal to considerably reduce the need for food during the winter. The burrow buffers extreme ambient temperatures, offers relatively stable climatic conditions, and protects against predators. Golden hamsters dig their burrows generally at a depth of 0.7 m (Gattermann et al., 2001). A burrow includes a steep entrance pipe [4-5 cm diameter], a nesting and a hoarding chamber and a blind ending branch for urination. Hamsters use their fore and hind legs as well as their snout and teeth for digging. Laboratory hamsters have not lost their ability

Comfortable Quarters for Laboratory Animals Reinhardt V, Reinhardt A (eds), 33-37. Animal Welfare Institute, Washington, DC 20007

to dig burrows (Kuhnen, 1986). In fact, they will do this with great vigor and skill if they are provided with the appropriate substrate (Editors' note).

Hamsters are nocturnal rodents who are active during the night, busily searching for food like fruits, vegetable, grains, worms and insects. They carry and hoard the food in large cheek pouches.

The golden hamster uses so-called flank organs [glands located in the hip region] for territorial marking. The territory is defended against conspecific intruders with overt aggression. Hamsters are vicious fighters and show little inhibition to inflict serious injuries. Females are bigger than males, and they are also more aggressive. Non-aggressive social contact among hamsters seems to be restricted to the brief instance of copulation and the period of lactation.

Hamster-adequate Housing

The environments of laboratory animals are usually designed on the basis of economic, hygienic, ergonomic and research aspects, with little consideration of animal welfare. Poor housing conditions can affect normal physiological responses and inhibit the expression of species-typical behaviors that are essential to safeguard the well-being and behavioral health of the animal.

Any housing system for hamsters should satisfy the physiological and ethological needs for resting, nest building, grooming, exploring, climbing, hiding, digging, searching for food, hoarding and gnawing.

Cage size

Legal minimum cage size specifications are not satisfactory because they are based on an individual animal's body weight, with more space being allocated to heavier—and hence older—animals than to lighter, younger animals. This simplistic formula fails to take into consideration that young, relatively light hamsters are highly motivated to engage in energetic exploratory and play activities requiring extra space. Keeping them in small cages results in chronic stress as measured in an impaired febrile response, i.e., increased baseline core temperature in response to the administration of a fever-inducing substance. Comparing the febrile response of hamsters housed in different cage sizes indicates that a hamster should not be kept on less than 800 cm^2 [124 in^2] floor space (Kuhnen, 1999a). To house hamsters in smaller cages may jeopardize sound interpretation of experimental results.

Breeding females should be kept in primary enclosures that provide no less than 1800 cm^2 [280 in^2]. The female hamsters should not be handled or disturbed from about 2 days before to 7 days after parturition.

The importance of cage space for hamsters can also be deduced from anatomical differences indicating a higher fitness [bigger lungs and bigger hearts] of hamsters housed in relatively large cages versus hamsters housed in small cages (Kuhnen, unpublished results). Housing hamsters in relatively large cages is not only directly beneficial for the animals but also indirectly by decreasing the interval of the extremely stressful cage cleaning procedure (Kuhnen, 1999a; Gattermann and Weinandy, 1996; Conn et al., 1990).

In the typical bipedal position (Figure 1) golden hamsters reach a height of about 16 cm. The primary enclosure should, therefore, have a minimum height of 17 cm [6.7 in].

Hamsters organize their quarters into separate functional areas: a sleeping nest with adjacent larder, a separate urination spot and an area for climbing and exploring.

Cage material and design

The caging material must be smooth, impervious to moisture and liquids, corrosion resistant, easy to clean and durable to withstand hamsters from gnawing through. Stainless steel impairs the hamster's visual orientation, may produce unpleasant noises and can create an uncomfortable ambiance due to its high thermal conductance, but it can offer the animal some shelter in the dark corners. Polycarbonate seems to be the better choice, because it allows the hamster to see what is going on outside the cage and enables care personnel to readily check the animal. Polycarbonate, however, does not offer the hamster any shelter. This drawback can be overcome with self-adhesive black foil patterns fixed on the outside surface of the cage walls and shelter-providing structural enhancement gadgets, such as sections of PVC tubing and small cardboard tubes or boxes (Figure 2).

Hamsters are very keen to get out of their cages. The cage lids must therefore fit securely to prevent escape. Golden hamsters are extremely good climbers and like to use the wire lids for gymnastics.

Hamsters show a strong preference for solid-floor cages with bedding over barren, wire-mesh floored cages (Arnold and Gillapsy, 1994). The animals should, therefore, be housed in obviously more comfortable solid-floor cages with bedding (cf., National Research Council, 1996).

Bedding

The bedding should allow digging and nest-building behavior, absorb urine, smells and toxins [particularly ammonia, which is very irritating to eyes and airways] and keep the hamster clean and dry. Dust-free woodchips or a combination of granulated softwood/filtered sawdust and dust-free

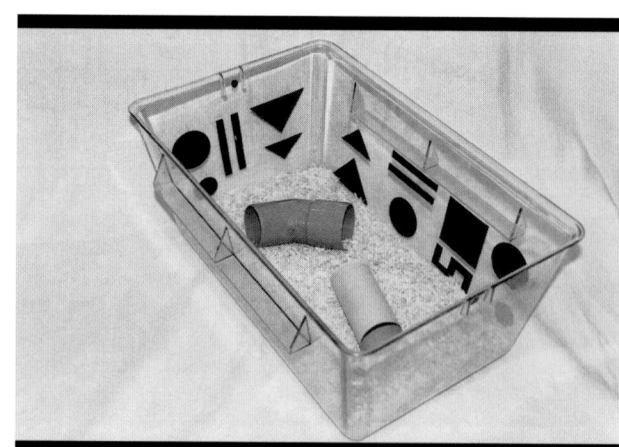

Figure 2. A polycarbonate cage with black graphics on two walls and structural enhancement gadgets.

woodchips are a good compromise between behavioral and hygienic needs. The change of bedding and/or cage induces a strong stress response in hamsters (Kuhnen, 1999a; Gattermann and Weinandy, 1996). In order to buffer some of the stress, the animal should always be allowed to retain some of the old, familiar nest material (Baumans, 1998).

Shelter and Nest

Hamsters are nocturnal animals who spend most of their time in a dark, secure burrow. Under laboratory housing conditions, direct, bright illumination must be avoided and crepuscular or dark retreat areas provided. They need to have access to a proper place of refuge and appropriate material and space to build a sleeping nest with adjacent larder. A simple, U-formed, opaque structure made of plastic material can serve as a suitable shelter, which is often used also as a nest site (Figure 3). In contrast to the closed nest box, the U-formed shelter permits care personnel to visually check the hamster without causing any disturbance. In order to construct a comfortable nest, the hamster should be provided with hay or wood-wool, paper tissue and a small tube of cardboard (Figure 3). For a hamster, nesting material is similarly important as is food and water, and he will readily work to get access to it (Jansen et al., 1969).

Supplemental food

To satisfy the foraging, food processing and hoarding drive of hamsters, natural food like grain, seeds, peanuts, carrots, root vegetables and pieces of apple should be distributed frequently on their bedding in addition to the standard food ration.

Enrichment devices

During the night hamsters have a very strong urge to actively interact with their physical environment. They should have access to different kinds of tubes [diameter ≥4 cm] stimulating explorative behavior (Figure 4) and indirectly increasing the useable floor space by providing additional wall contact and refuge. The provision of tubes can be an effective means to resolve aggressive behavior problems such as growling, hissing, aggressive posturing toward humans, destruction of water bottle rubber stoppers and attacking objects introduced into the cage (McClure and Thomson, 1992). The value of exercise wheels is still debated. The hamsters' intensive running in the wheel may simply be a locomotor stereotypy. The demand for a running wheel decreases with increased cage size, suggesting that rather than fostering stereotypical wheel-running in small cages, hamsters should be allowed to express their urge to move around in sufficiently spacious cages.

Hamsters should always have access to pieces of soft wood to satisfy their need for gnawing and prevent overgrowth of their teeth.

The positive effects of enriched housing on problem-solving tests is well documented since the experiments of Hebb (1947) in rats. In hamsters, environmental enrichment improves the febrile response to restricted cage space without a side-effect on the dispersion of the measured values (Kuhnen, 1999a). Behavioral tests indicate less fear and a better orientation (Kuhnen, 1999b) and improved spatial

Figure 3. A U-formed opaque shelter with hay, paper tissue and a tube of cardboard serve a golden hamster as appropriate materials to construct a sheltered, comfortable nest.

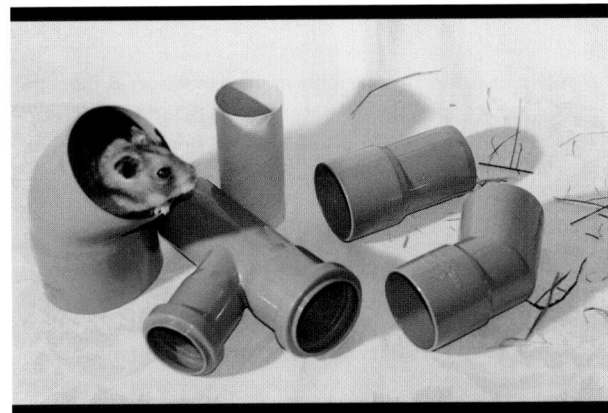

Figure 4. Plastic tubes are perfect enrichment devices for golden hamsters.

perception and discrimination (Thinus-Blanc, 1981) in golden hamsters housed in enriched cages.

Single-housing versus social-housing

Unlike all other animals commonly found in research facilities, golden hamsters are a strictly solitary species. Living in groups is artificial for mature hamsters. Enforced group-housing will result in severe and chronic social stress (Meisel et al., 1990; Zimmer and Gattermann, 1996) and a higher rate of wounding (Arnold and Estep, 1990). It is possible to keep immature male siblings in unisexual groups without unduly jeopardizing the animals' safety.

When in oestrus, the female hamster usually gets agitated and accepts copulation from any sexually competent male. She will exhibit lordosis as a clear sign of receptivity. If not receptive, females are likely to attack and possibly kill males. It is favorable to set a receptive female into the territorially marked cage of a male to give the male the advantage of being on his own turf. The female—if fully receptive—will be less inclined to be aggressive under this circumstance. A

Figure 5. Golden hamsters readily overcome their instinctive fear of humans when they are regularly handled with gentleness.

male who is older, heavier and more experienced than the female has a good chance of not being beaten up. In consideration of the male's safety, the female has to be removed right after mating and returned to her own cage.

Assessment of Well-being

Daily visual, unobtrusive inspections of caged hamsters are important to notice pain, suffering or injuries in time so that prompt treatment can be initiated. The following signs are indicators that a hamster is not well:

- General: little or no interest in any food, loss of weight, change of circadian rhythm;
- Eyes: reddening of the conjunctiva, discharge, adhered eyelids;
- Behavior: increased aggression, depression, indolence, decreased exploration;
- Posture: curled posture in the awake hamster, rigid standing, stiff gait;
- Appearance: wet tail, diarrhea, loss of hair, scaled skin, blunt fur.

A healthy golden hamster will typically go through the following ritual when waking up: stretching and yawning with fully opened mouth followed by grooming.

Hamster-adequate Handling

Hamsters are nocturnal animals who instinctively avoid being touched by man. It is therefore important to gently wake a hamster up before manipulating him or her during daytime hours. This not only helps to prevent data-influencing stress reactions, but a startled hamster is frightened and may react aggressively to the disturbing intruder. Hamsters are readily tamed and habituated to gentle handling (Figure 5). Regular gentle handling of a research hamster is no waste of time but will pay off in more reliable research data.

References

Arnold CE, Estep DQ 1990. Effects of housing on social preference and behaviour in male golden hamsters (*Mesocricetus auratus*). Applied Animal Behaviour Science 27, 253-261

Arnold CE, Gillapsy S 1994. Assessing laboratory life for golden hamsters: social preference, cage selection, and human interaction. Lab Animal 23, 34-37

Baumans, V 1998. Enrichment bei Labormäusen: Eine Sache für Mäuse und Menschen. In Tierlaboratorium 21 Juhr N-C, Hiller HH, Scharmann W (eds), 59-62. Zentrale Universitätsdruckerei, Berlin, Germany

Conn CA, Borer KT, Kluger MJ 1990. Body temperature rhythm and response to pyrogen in exercising and sedentary hamsters. Medicine and Science in Sports and Exercise 22, 636-642

Gattermann R, Weinandy R 1996. Time of day and stress response to different stressors in experimental animals. Journal of Experimental Animal Science 38, 66-76

Gattermann R 2000. 70 Jahre Goldhamster in menschlicher Obhut—wie groß sind die Unterschiede zu seinen wildlebenden Verwandten. In Tierlaboratorium 23 Juhr NC, Hiller HH, Scharmann W (eds), 86-99. Zentrale Universitätsdruckerei der Freien Universität, Berlin, Germany

Gattermann R, Fritzsche P, Neumann K, Al-Hussein I, Kayser A, Abiad M, Yakti R 2001. Notes on the current distribution and ecology of wild golden hamsters (*Mesocricetus auratus*). Journal of Zoology 254, 359-365

Hebb DO 1947. The effect of early experience on problem-solving at maturity. American Psychologist 2, 306-307

Jansen PE, Goodman ED, Jowaisas D, Bunnell BN 1969. Paper as a positive reinforcer for acquisition of a bar press response by the golden hamster. Psychonomic Science 16, 113-114

Kuhnen G 1986. O_2 and CO_2 concentrations in burrows of euthermic and hibernating golden hamsters. Comparative Biochemistry and Physiology 84A, 517-522

Kuhnen G 1999a. The effect of cage size and enrichment on core temperature and febrile response of the golden hamster. Laboratory Animals 33, 221-227

Kuhnen G 1999b. The effect of housing conditions on the results of behavioural tests in golden hamster. Zoology 102 (Supplement 2), 84

McClure DE, Thomson JI 1992. Cage enrichment for hamsters housed in suspended wire cages. Contemporary Topics in Laboratory Animal Science 31, 33

Meisel RL, Hays TC, Del Paine SN, Luttrell VR 1990. Induction of obesity by group housing in female syrian hamsters. Physiology and Behavior 47, 815-817

National Research Council 1996. Guide for the Care and Use of Laboratory Animals, 7th Edition. National Academy Press, Washington, DC
Full Text: http://www.nap.edu/readingroom/books/labrats/

Thinus-Blanc C 1981. Volume discrimination learning in golden hamsters: effect of the structure of complex rearing cages. Developmental Psychobiology 14, 397-403

United States Department of Agriculture 2000. <u>Animal Welfare Report–Fiscal Year 2000</u>. U.S. Department of Agriculture–Animal Care, Riverdale, MD
Full Text: http://www.aphis.usda.gov/ac/awrep2000.pdf

Zimmer R, Gattermann R 1996. Der Einfluß von Haltung und Rang auf die Nebennierenaktivität männlicher Goldhamster (*Mesocricetus auratus*). <u>Zeitschrift für Säugetierkunde</u> 61 (Sonderband), 74-75

Dr. Gernot Kuhnen is a lecturer at the Physiological Institute of the University of Giessen, Germany, and head of the 'Büro für Thermophysiologie.' He is a biologist and has worked for many years with hamsters studying hibernation, thermoregulation and housing systems.

Comfortable Quarters for Guinea-pigs in Research Institutions

Viktor Reinhardt

Animal Welfare Institute, PO Box 3650, Washington, DC 20007, USA

Species-characteristics

Guinea-pigs are domesticated, conspicuously docile, social rodents originating from South America. Emitting squeaky contact sounds like little pigs is the explanation for their misleading name. They live in small groups of five to ten individuals. Even though they do not groom one another, guinea-pigs seek each other's bodily contact during periods of rest. Scent marking with urine squirted on the coat of another partner, and scent marking with secretions from perineal and supracaudal glands rubbed on the substratum reflect the animals' relative social status and social roles within the group. Strange individuals are recognized by the absence of group-characteristic scents (Reinhardt, 1971).

The young are born after a relatively long gestation period of about 66 days. Unlike other rodents, guinea-pigs do not construct nests. In fact, the newborn are so precocial that they do not need a nest. They look like small-sized adults and start nibbling and eating solid food on the day of their birth. Young guinea-pigs—unlike adults—are very active and engage in exuberant running-and-hopping games alone or with other peers. However, the stage of infantile gamboling is very brief. Young females may successfully breed when they are only three weeks old and give birth to their first litter at the age of three months. Young males engage in sexual courtship activities also in their third week, and by doing so become sexual competitors for the boss male of the group who will target them with persistent chasing. In this way, young males gradually become sexually and behaviorally inhibited unless they are removed from the group (Reinhardt, 1971).

Guinea-pigs neither compete over food nor do they hoard food. This leaves little reason for aggressive disputes. Females never engage in fighting and only rarely do they have harmless squabbles with each other. They get along with each other so well that they even practice communal nursing. In sharp contrast to rats, mice, gerbils and hamsters, guinea-pig mothers do not seem to care much about their own offspring. They neither groom their young nor will they defend them if need arises. If several females have newborns, a naïve observer is not able to tell from the animals' behavior which offspring belongs to which female, because

the mothers treat all the young equally and the young suckle from any lactating female. Guinea-pig mothers nurse for no longer than three weeks. During that period of lactation they are tolerant nurses for all—not only for their own—infants of the group. It is the lactating female—not the young—who sets the timetable for nursing. If she feels ready to nurse, she will get restless and walk back and forth thereby attracting the attention of all infants, who will gather and follow until the "nurse" squats on her favorite location and tolerates being suckled. Once she has started, the nurse will butt away any straggler even if it happens to be one of her own progeny. After about ten minutes, she will abruptly walk away leaving a heap of perplexed young. Even though females seemingly do not develop a bond with their offspring, the presence of suckling young makes them aggressive toward strange females (Reinhardt, 1971).

Comfortable Quarters for Laboratory Animals Reinhardt V, Reinhardt A (eds), 38-42. Animal Welfare Institute, Washington, DC 20007

Males are inhibited to show any kind of aggression—including threats—against females, but they viciously fight with each other in the presence of oestrus females. Usually, there is one male who monopolizes such females; his strong dominance keeps all other males in a state of social distress most of the time. In order to prevent attacks, subordinate males behave like females and stop emitting male-typical pheromones. The dominant male treats such "frustrated" subordinates like females and displays courtship rather than aggressive behaviors (Reinhardt, 1971).

Adult guinea-pigs measure up to 30 cm in length. For the stretching posture they need approximately 3 cm additional horizontal space to allow free expression of this comfort behavior (Figure 1). They are poor diggers but greatly enjoy burrowing in hay. Vocalization plays an important role in their social and sexual behavior. There is always some purring, whistling, squeaking or teeth-chattering to be heard in a guinea-pig room, and a friendly caretaker can count on receiving a noisy welcome. Like most other rodents, guinea-pigs are susceptible to noise stress (Anthony et al., 1959).

Guinea-pigs are relatively heavy rodents—with adults weighing about 1 kg—and are, therefore, prone to developing pressure sores and pododermatitis on wire mesh floors (cf., Fullerton and Gilliatt, 1967). They are distinctively quadrupedal animals who usually keep all four feet on the ground and, unlike most other rodents, do not show a bipedal orientation stance. Therefore, relatively little head room—approximately 20 cm—is required when keeping them in enclosures. There is no need for a cover because the animals are poor climbers and will normally make no serious attempts to escape over walls that are only 30 cm high (Reinhardt, 1971).

Figure 1. Caged guinea-pigs must be provided with sufficient horizontal space so they can stretch freely in a species-typical manner.

"Rodents appear to prefer sheltered areas of the cage, especially if those areas have decreased light and height. Providing such a confined space within a cage might be one way to enrich the environment of rodents" (National Research Council, 1996, p. 48). The provision of such protected, safe space is a basic requirement to assure data collection of animals who are not unduly stressed by their living environment. Guinea-pigs instinctively avoid open surfaces that expose them full-view to potential predators. They will always keep close to walls and shun the empty, unprotected central area of an enclosure (White et al., 1989). To conclude from this that "social groups of rodents do not use all the available space recommended in current guidelines and probably do not require it for well-being" (National Research Council,

Figure 2. Guinea-pigs need the social group environment for their well-being. Bedding of dust-free shavings, supplemented with hay is a basic, yet effective form of environmental enrichment and feeding enrichment for guinea-pigs (photo by James Love, University of British Columbia, Vancouver, Canada).

1996, p. 48) is a misleading generalization that could be twisted in such a way that the animals are granted only the bare minimum space without option to move around at all.

Addressing Guinea-pig Specific Characteristics in the Research Institution

"All who care for or use animals in research, teaching, or testing must assume responsibility for their well-being....A good management program provides the environment, housing, and care that...minimizes variations that can affect research results" (National Research Council, 1996a, pp. 1 & 21) and hence, reduces the number of research subjects needed to achieve statistically significant results. In the United States approximately 500,000 guinea-pigs were used in research during the year 2000 (United States Department of Agriculture, 2000).

Guinea-pigs need the **social environment** to guarantee their behavioral health, safeguard their physiological well-being (Sachser and Lick, 1991; Fenske, 1992), and assist them to cope with circumstances of confinement (Canadian Council on Animal Care, 1993). Compatible group-housing should, therefore, be the standard arrangement for them in the research laboratory (cf., Brain et al., 1994; Council of Europe, 2000a; Figure 2).

Animals living in groups should be provided a floor area of no less than 1200 cm^2 per breeding female, and no less then 750 cm^2 per weaned, non-breeding animal. A box with an access hole makes the central area of the enclosure a **place of refuge** for the animals thereby improving their well-being and, at the same time, increasing the usable floor space (Figure 2). Such a refuge is an attractive location to sleep together and for pregnant females to give birth. Provisioning the box with a sliding door is an elegant way of facilitating the capture of the animals for cage cleaning: The whole group is trapped in the box, which is then simply lifted out of the pen during the cleaning procedure. A removable top of the box allows the capture of individual animals (Gray, 1988).

To **minimize social tensions** arising from the presence of several mature males, and from overcrowding, it is recommended to keep only one mature male with 3-6 females and remove the naturally weaned young at the age of three weeks. The adolescents can then be housed in same-sex groups without risk, provided that male groups have no visual and no olfactory contact with female groups. The mere exposure to the smell of female urine will turn even the most compatible males into fractious enemies who will no longer tolerate each other's presence (Reinhardt, 1971). Exchanging the mature breeding male of a group with another male constitutes no problem. The females will accept him and there will be no overt aggression. Strange females can be introduced into an existing group without causing social turmoil (Raje and Stewart, 2000) as long as none of the animals is nursing. When they lactate, females do not tolerate strange females (Reinhardt, 1971). It is, therefore, good advice to introduce females into an existing group only when none of the animals is lactating. An individual animal who has been removed from a group can be re-introduced

Figure 3. Guinea-pigs love fresh, leafy vegetables... and the person who feeds them.

without triggering xenophobic aggression, provided that he or she has not been scent-marked by another conspecific.

If a medical condition requires temporary **single-caging**, the floor area should be no less than 35 x 35 cm [1225 cm^2] in order to provide the space needed for an adult animal to stretch and turn around freely, and for a young animal to engage in locomotor play activities. If a research protocol requires temporary single-caging the floor space must be doubled [35 x 70 cm; 2450 cm^2] to allow the placement of a covered refuge area, such as a hiding box. Singly caged guinea-pigs must never be kept isolated, but provision must be made that they have visual, auditory and olfactory contact with others of their own kind (cf., Fenske, 1992; Canadian Council on Animal Care, 1993).

Multiple-tier caging systems are not recommended because the shade cast from upper tiers on lower tiers makes it impossible to assure that "lighting shall provide uniformly distributed illumination" (United States Department of Agriculture, 1995a, p. 53; cf., Bellhorn, 1980; Clough, 1982). **Uniform lighting** for all animals, however, is a fundamental condition of scientifically valid research methodology (American Medical Association, 1992) assuring that no more than the minimum number of guinea-pigs are used to obtain statistically significant research results.

Guinea-pigs do well on a commercial pelleted diet supplemented daily with fresh produce and hay. "When good quality **hay** is supplied the consumption of the more expensive pelleted diet is reduced and, by their vocalization when they realize that the hay is about to be replenished, the animals clearly indicate the great pleasure they obtain from eating it and burrowing in it" (Sutherland and Festing, 1987, p. 401). Guinea-pigs must engage in regular gnawing behavior to prevent overgrowth of their front teeth. Hard food pellets, carrots and softwood sticks (Scharmann, 1991) are suitable to meet this need.

Regular distribution of food treats, such as fresh, leafy produce (Figure 3) and gentle **handling** help guinea-pigs to overcome their fear of personnel. They should be picked up gently with both hands, one firmly around the shoulder and the other supporting the hindquarters. "Animal care staff are expected, at all times, to have a caring and respectful attitude towards animals in their care, and to be proficient in their handling" (Council of Europe, 2000b, p. 27). Proper handling depends on the investigator rather than on the subject. "Investigators should [always] consider that procedures that cause pain or distress in human beings may cause pain or distress in other animals" (American Association for Laboratory Animal Science, 1997, p. 51). Nervous, impatient or even callous investigators startle and distress guinea-pigs, rendering research data collected from such animals virtually useless. Guinea-pigs should be handled "as expeditiously and carefully as possible in a manner that does not cause trauma, overheating,…behavioral stress,…or unnecessary discomfort" (United States Department of Agriculture, 1995a, pp. 21-22).

The rooms in which guinea-pigs are housed have to be **quiet** to avoid stress responses triggered by intense and/or chronic noise (cf., Anthony and Harclerode, 1959).

With the exception of specific short-term experimental protocols, guinea-pigs should always be kept on **solid floor with bedding** (cf., National Research Council, 1996a). "When grid or perforated floors are used, a solid resting area must be provided" (Council of Europe, 2000a, p. 5) that is sufficiently large to allow all animals to lie on it simultaneously. Bedding of dust-free shavings from seasoned soft wood, supplemented daily with high-quality hay, should be regarded as a basic form of environmental and feeding enrichment (Figure 2). Regularly—at least twice a week—changed bedding is the best guarantee of a hygienic cage environment.

Conclusion

To make the quarters for guinea-pigs in research institutions "comfortable" the following provisions are recommended:

- Compatible social housing;
- Protected refuge area(s);
- Solid resting area(s);
- Dust-free wood shavings bedding;
- Commercial pelleted diet supplemented daily with hay and fresh produce;
- Regular distribution of food treats to foster a positive human-animal relationship;
- Gentle-and-firm handling.

References

American Association for Laboratory Animal Science 1997. AALAS policy on the humane care and use of laboratory animals. Contemporary Topics in Laboratory Animal Science 36, 51

American Medical Association 1992. Use of Animals in Biomedical Research—The Challenge and Response—An American Medical Association White Paper. AMA. Group on Science and Technology, Chicago, IL

Anthony A, Harclerode JE 1959. Noise stress in laboratory rodents. II: Effects of chronic noise exposure on sexual performance and reproductive function of guinea pigs. Journal of the Acoustical Society of America 31, 1437-1440

Anthony A, Ackerman E, Lloyd JA 1959. Noise stress in laboratory rodents. I. Behavioral and endocrine response of mice, rats and guinea pigs. Journal of the Acoustical Society of America 31, 1430-1436

Bellhorn RW 1980. Lighting in the animal environment. Laboratory Animal Science 30, 440-450

Brain PF, Büttner D, Costa P, Gregory JA, Heine WOP, Koolhaas J, Militzer K, Ödberg FO, Scharmann W, Stauffacher M, Baumans V, Poole TB, Sachser N, Whittaker D 1994. Rodents. In The Accommodation of Laboratory Animal in Accordance with Animal Welfare Requirements. O'Donoghue PN (ed), 1-14. Bundesministerium für Ernährung, Landwirtschaft und Forsten, Bonn, Germany

Canadian Council on Animal Care 1993. Guide to the Care and Use of Experimental Animals, Volume 1, 2nd Edition. Canadian Council on Animal Care, Ottawa, Canada
Full Text: http://www.ccac.ca/guides/english/toc_v1.htm

Clough G 1982. Environmental effects on animals used in biomedical research. Biological Reviews 57, 487-523

Council of Europe—Working Party for the Preparation of the Fourth Multilateral Consultation of Parties to the European Convention for the Protection of Vertebrate Animals Used for Experimental and Other Scientific Purposes (ETS 123) 2000a. Revised Proposals for Rodents & Rabbits. Council of Europe, Strasbourg, France

Council of Europe—Working Party for the Preparation of the Fourth Multilateral Consultation of Parties to the European Convention for the Protection of Vertebrate Animals Used for Experimental and Other Scientific Purposes (ETS 123) 2000b. Proposal II (General part of Appendix A). Council of Europe, Strasbourg, France

Fenske M 1992. Body weight and water intake of guinea pigs: influence of single caging and an unfamiliar new room. Journal of Experimental Animal Science 35, 71-79

Fullerton PM, Gilliatt RW 1967. Pressure neuropathy in the hind foot of the guinea pig. Journal of Neurology, Neurosurgery and Psychiatry 30, 18-25

Gray G 1988. Guinea pigs. Humane Innovations and Alternatives in Animal Experimentation 2, 48-49
Full Text: http://www.awionline.org/lab_animals/biblio/hiaa-88.html

National Research Council 1996a. Guide for the Care and Use of Laboratory Animals, 7th Edition. National Academy Press, Washington, DC
Full Text: http://www.nap.edu/readingroom/books/labrats/

National Research Council 1996b. Laboratory Animal Management Rodents. National Academy Press, Washington, DC
Full Text: http://www.nap.edu/readingroom/records/0309049369.html

Raje SS, Stewart KL 2000. Group housing female guinea pigs. Lab Animal 29(8), 31-32
Full Text: http://www.awionline.org/lab_animals/biblio/la29-8gp.html

Reinhardt V 1971. Soziale Verhaltensweisen und soziale Rollen des Hausmeerschweinchens [Social behavior and social roles of guinea-pigs]. Dissertationsdruck Novotny, Söcking/Starnberg, Germany

Sachser N, Lick C 1991. Social experience, behavior and stress in guinea pigs. Physiology and Behavior 50, 83-90

Scharmann W 1991. Improved housing of mice, rats and guinea-pigs: a contribution to the refinement of animal experimentation. Alternatives to Laboratory Animals [ATLA] 19, 108-114

Sutherland SD, Festing MFW 1987. The guinea-pig. In The UFAW Handbook on the Care and Management of Laboratory Animals, Sixth Edition Poole TB (ed), 393-410. Churchill Livingstone, New York, NY

United States Department of Agriculture 1995a. Regulations under the Animal Welfare Act as Amended (7 USC, 2131-2156). 9 CFR Ch. 1 (1-1-95 Edition)
Full Text: http://www.access.gpo.gov/nara/cfr/waisidx_00/9cfr2_00.html

United States Department of Agriculture 1995b. Standards under the Animal Welfare Act as Amended (7 USC, 2131-2156). 9 CFR Ch. 1 (1-1-95 Edition)
Full Text: http://www.access.gpo.gov/nara/cfr/waisidx_00/9cfr3_00.html

United States Department of Agriculture 2000. Animal Welfare Report—Fiscal Year 2000. U.S. Department of Agriculture—Animal Care, Riverdale, MD
Full Text: http://www.aphis.usda.gov/ac/awrep2000.pdf

White WJ, Balk MW, Lang CM 1989. Use of cage space by guinea pigs. Laboratory Animals 23, 208-214

Viktor Reinhardt is Laboratory Animal Advisor to the Animal Welfare Institute in Washington, DC. He studied the social behavior and social roles of captive guinea-pigs for his doctoral dissertation.

Comfortable Quarters for Rabbits in Research Institutions

Boers K., Gray G., Love J., Mahmutovic Z., McCormick S., Turcotte N., Zhang Y.
Animal Care Centre, University of British Columbia, Vancouver BC, Canada V6T 1W5

In the past, laboratory rabbits have usually been kept singly in small cages that provide neither social nor environmental enrichment, and frequently the cages have been too small to permit some normal behaviors such as sitting up on the hind legs, hopping, digging and hiding. Stereotypical bar licking or chewing, pawing at the corners of the cage, psychogenic polydipsia and excessive self-grooming—resulting in the development of trichobezoars—are frequently seen in such animals and are recognized as indicators of reduced well-being. Single-caged rabbits often look unhealthy and depressed, sitting in a hunched position for hours on end. The extreme boredom induces some animals to overeat, others to undereat, leading to obesity and severe weight loss, respectively (Gunn-Dore, 1997). In addition, there are serious clinical problems associated with the small cage. Degenerative changes of the lumbar spine and femoral head have been attributed to the lack of basic locomotor activity in the small, conventional cages (Wieser, 1986). It is questionable if statistically reliable and scientifically meaningful research data can be obtained from animals kept under such inadequate housing conditions.

There have been several reports of improved housing for laboratory rabbits and most have a base in the behavior of the wild European rabbit (*Oryctolagus cuniculus*; Morton et al., 1993; Love, 1994; Wemelsfelder, 1994; Gunn-Dore, 1997). Detailed descriptions of this behavior are available from the study of rabbits in their natural environment and in semi-wild conditions (Mykytowycz, 1958; Kraft, 1978; Lehmann, 1991). While it is usually impossible to accommodate all behaviors seen in the wild, at least basic behavioral needs can readily be addressed in the laboratory setting.

Wild rabbits are social animals who interact with each other whether they live in large groups or small groups. Aggression among females is limited although dominance hierarchies are formed, and females with young will chase other rabbits away from their nests. Aggression among males increases as they approach puberty and consists mainly of chasing, with one rabbit trying to get out of the sight of the other. Amicable interactions (e.g., mutual grooming, lying close together) are usually seen only in the sexual context between a buck and a doe. Female/female amicable interactions occur under laboratory conditions in the absence of males. Both sexes may participate in scent marking of inanimate objects. Young rabbits sport and play with each other and with inanimate objects. In the wild, rabbits dig burrows to hide and nest in, and they dig for the roots of plants. In the laboratory, rabbits will dig for no obvious reason, indicating that they are highly motivated to engage in this activity.

This very brief review of the rabbit's species-typical behavior provides us with some indicators for the development of comfortable housing. We are usually restricted by the available space and by some experimental limitations. Pregnancies are undesirable, except in the breeding colony. Mature males and females can, therefore, not be kept together.

The most suitable quarters for rabbits allow for social interaction and provide physical substrate for digging, playing and hiding. Several authors have described housing systems that provide these needs, both for breeding colonies and experimental animals. There have also been descriptions of short-term, single-housing systems that attempt to address as many of the needs as possible.

The Rabbit Pen

Rabbits are gregarious animals and, therefore, should be housed in compatible groups (Stauffacher et al., 1994). Each rabbit is provided with substantially more living space and hence has much better opportunities for exercising in a group-pen than in a single-cage. **The quality of life of group-housed rabbits is significantly improved,** even of individuals who rank low in the social hierarchy, compared to those kept in solitary confinement (Held, 1996; Batchelor, 1999). Group members spend an average of 79% of the time in close proximity with others (Gunn and Morton, 1993). Behavioral disorders, which typically occur in single-caged rabbits, are virtually absent in group-housed rabbits (Loeffler et al., 1991; Podberscek et al., 1991; Love, 1994; Krohn et al., 1999; Held et al., 2001). Compatible group-housing does not significantly affect stress-sensitive variables and infectious disease susceptibility (Love and Hammond, 1991; Gunn-Dore, 1997; Turner et al., 1997).

Figure 1. Rabbits have a preference for straw as bedding (photo by Debbie Gunn-Dore, 1997).

Figure 2. The resting boards of the converted dog runs provide a comfortable place for the rabbits to sit on and to hide under.

Female rabbits are generally compatible with each other. Given a choice, they prefer to be in the company of another doe than living alone (Brooks et al., 1993). Housing female rabbits in pairs or groups not only allows them to express their social needs, but it also makes them less susceptible to stress than single-caged does. The company of other rabbits has an emotionally protective effect during stressful situations.

Male rabbits develop a biological intolerance of other males when reaching sexual maturity at the age of 12-14 weeks. Young males can and should be housed in a social setting until that time, but they have to be separated from other males thereafter to prevent injuries resulting from fighting. Castration prior to puberty can resolve this problem (Love and Hammond, 1991). Single-housed bucks should not live in social isolation, but they should be able to see and possibly touch and smell other rabbits without being able to engage in fighting.

Rabbits, like all social animals, develop dominance-subordination relationships that are a prerequisite for a harmonious group life. Removing or replacing an adult group member inevitably disrupts these relationships and may lead to serious aggressive disputes. **It is very important to keep the composition of a group stable.** Individual animals who have to be temporarily separated for experimental or clinical reasons should always be housed in such a way that they can maintain visual contact with the group. This ensures that they will be readily recognized and accepted as familiar members of the group upon returning. It is often said that a rabbit who has undergone a surgical procedure should be isolated so that other rabbits don't abuse him or her and nibble at the sutures. We have found that this rarely occurs. It is our experience that rabbits lie down beside a group member who is returning from a surgery, and that this extra warmth and comfort hastens the recovery process.

The primary enclosure of a rabbit group should be large enough to allow three hops in one direction. A fully grown New Zealand White rabbit will move forward 1.5 to 2.0 m in three such hops (Love, 1994). **Hence, the pen should mea-**

Figure 3. Rabbits make use of shelves to access the vertical dimension of their pen.

sure at least 2 m in one direction. If more than two adult rabbits of the weight category 4-6 kg are housed together, the minimum floor area of the primary enclosure should be 2 m^2 for up to four animals, increasing by 0.45 m^2 for each additional adult rabbit (Gunn-Dore, 1997). The height of the pen should be not less than 1.20 m to prevent the rabbits from leaping out. If a wire mesh cover is used to keep the animals in, it must be at least 75 cm above the floor to allow adult rabbits to sit in the lookout posture.

The rabbit pen should be provisioned with woodchip litter or preferentially with **shredded paper or straw** bedding (Figure 1). When given the choice, rabbits prefer straw or shredded paper and avoid sawdust or wood shavings (Turner

Figure 4. Rabbits relaxing in their indoor-outdoor pen.

et al., 1992). **Hay** must be provided for foraging and nest-building. There must be nest boxes for breeding females, designed in such a way that they make it impossible for littering does to see each other and trigger infanticide behavior. **Shelves** should make the vertical dimension accessible and offer comfortable resting and refuge places (Figure 2 & 3). **Wooden sticks** and tree branches are suitable to promote gnawing behavior. Rabbits will spend about 20% of the time gnawing such objects (Stauffacher, 1992). Cardboard boxes, plastic crates and/or sections of 18-inch PVC pipe should be available as substitute **burrows** and "safe" places to retreat in fear-provoking situations or during social conflicts. At least one wall of the enclosure should consist of wire mesh so that the animals can overlook their surroundings and see approaching personnel (Figures 1 & 3).

The rabbits at our facility are housed in pens that were originally designed for dogs (cf., Tamburrino et al., 1999; Figure 2 & 4). Each pen holds six to eight animals. There is an indoor section with a resting board and an outdoor section. The rabbits move freely from one area to the other. The indoor section measures approximately 1.5 x 1.7 m, the outside section 3.5 x 1.7 m. We have noted that our rabbits like to explore, and that they do not mind climbing. They often sit on the resting boards (Figure 2) and when given the opportunity, will climb much higher and seem quite relaxed about it (Figure 3). The outdoor run allows the rabbits to indulge in "fast running," an activity that we frequently observe, particularly in young animals. A rabbit runs quickly to one end of the pen, stops and then runs quickly to the other. This may be repeated several times. We have never observed a special reason for this exercise, other than that the animals obviously enjoy it.

Establishing a New Group of Rabbits

Group sizes of four to eight adult rabbits work well if the groups are to remain together for a long time. Larger groups of subadult rabbits may be maintained for short periods of time. It is good advice to establish a new group with **young animals** who have not reached puberty. Group members should be of the same age and sex, but it is not necessary that they are littermates.

Group-housing rabbits who have been previously single-caged for more than six months is not recommended. Such animals will be extremely fearful, will lack proper motor coordination resulting from long-term hypoactivity and will be prone to injuries and fractures due to weakness in the bone structure (Drescher and Loeffler, 1991; Rothfritz et al., 1992; Gunn-Dore, 1997). **Pair-housing** them in double-cages minimizes these risks while offering a more species-adequate, social environment (Bigler and Oester, 1994). Pair-housing is recommended for immature rabbits, adult females and castrated males (Huls et al., 1991; Stauffacher, 1992; Bigler and Oester, 1994; Raje and Stewart, 1997). Mature bucks cannot be kept in pairs because of the serious risk of injurious aggression.

The Rabbit Cage

Under **exceptional circumstances**—such as research studies requiring urine collection—a rabbit may have to be single-housed for a limited period of time. Provision must be made that such an individual animal is not visually isolated from other rabbits and that his or her cage is sufficiently sized to allow normal postural adjustments with freedom of movement (United States Department of Agriculture, 1991) and is adequately enriched to relieve boredom.

An adult rabbit is about 75 cm tall when sitting in the rabbit-typical lookout posture (Figure 5) and approximately 80 cm long when resting in rabbit-typical lateral sternal recumbency (Figure 6). **The primary enclosure of single-**

Figure 5. An adult rabbit is about 75 cm tall when sitting in the rabbit-typical lookout postion (photo by Debbie Gunn-Dore, 1997).

housed animals should, therefore be at least 75 cm high and no less than 80 cm long. It should be 68 cm wide to allow the animal to comfortably turn around and change postures (Gunn-Dore, 1997).

Each cage should be provisioned daily with high-quality ***hay*** to promote the expression of foraging, playing, investigating and nesting behavior. The hay should be placed on the top of the cage so that the animal can spend some extra time retrieving it through the bars. There should also be at least one ***wooden stick*** [length approximately 10 cm; diameter approximately 2.5 cm] or other rabbit-suitable enrichment gadgets, such as brass wire balls triggering species-typical gnawing, chin-marking and playing (Huls et al., 1991; Gunn-Dore, 1999). Gnawing sticks have been used for a 2-year test period as effective enrichment objects for single- and pair-housed rabbits without noticeable hygiene and health problems (Brooks et al., 1993). It is a general idea at some facilities that rabbits need gnawing sticks to prevent their teeth from getting too long (Lindfors, 1997).

Single-caged rabbits who have access to hay and other enrichment objects show a reduction in stereotypical behaviors and a marked increase in their overall activity, relative to animals kept in barren cages (Gunn-Dore, 1997; Berthelsen and Hansen, 1999). Hay has proven to be particularly effective in reducing behavioral disorders and giving individually housed bucks something to do (Lindfors, 1997). The single-housed rabbit also needs a "safe" refuge to hide in alarming situations. A section of a PVC tube can serve as a substitute ***burrow*** meeting this requirement.

Cages should be designed in such a way that the rabbits are not restricted to grid or wire flooring—which is uncomfortable for the animals and very often results in sore hocks [ulcerative pododermatitis] (Kraus and Weisbroth, 1994)—but that they also have access to a ***raised solid-floor area***. This raised area offers a choice of resting sites, light gradients and a stimulus for exercise (Stauffacher, 1993; Gerson, 2000). The cages should be arranged at waist-height for easy access and cleaning. Multi-tier caging systems are not recommended because they do not allow the provision of uniformly distributed illumination (United States Depart-

Figure 6. An adult rabbit is approximately 80 cm long when resting in typical rabbit-fashion (photo by Debbie Gunn-Dore, 1997).

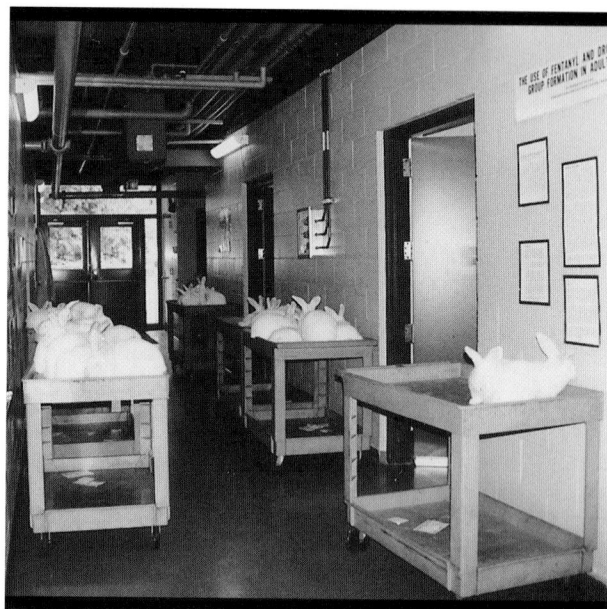

Figure 7. Unsedated rabbits waiting to have blood samples taken. The rabbits are accustomed to travelling to and from their pens on these carts.

ment of Agriculture, 1991), a prerequisite to avoid variability of research data resulting from variable illumination in the cages (Bellhorn, 1980; Clough, 1982).

The Animal Care Technician's Role in Providing a Stress-free Environment for Rabbits

Although comfortable housing is important for the rabbits, much of the effort would be wasted if the other activities surrounding the rabbits were not also comfortable and non-stressful. In this respect, the animal care technician plays a vital role. The following are examples where technician/rabbit interactions are important.

Group-housed rabbits must be caught with a minimum of chasing. We can make use of the rabbit's natural tendency to hide when startled. In our case, the rabbits duck under the resting board (Figure 2) where they may be identified, picked up and handled in a gentle and skillful manner. Any dark hiding place will serve the same purpose, but a quiet, smooth approach is required. ***It is important not to startle the animal in his or her hiding place.*** Once the animals are used to being picked up, they may not even hide from a technician they know well. The anticipation of what is to happen after being caught plays a major role in the rabbit's behavior. Procedures carried out with the rabbits should be as free of stress as possible. Rabbits who are used to being treated with compassion and professional skill will not panic in anticipation of procedures (Figure 7). Carefully bundling a rabbit in a blanket and gently covering his or her eyes with a towel usually has a calming effect, even on a very agitated animal.

The traditional rabbit restrainer for taking blood

samples is unnecessary if you provide good analgesia and some gentle handling. Blood sampling is least stressful if the subject is given a sedative and an analgesic. The added advantage is that the arteries and veins are dilated, making it easier to take the samples. Local anesthetics [e.g., EMLA cream] may serve the same purpose.

Rabbits have the potential of learning to cooperate rather than resist during procedures. It has been documented that they can easily be trained to cooperate during oral drug application, thereby avoiding the stressful gastric intubation procedure. The animals would stand with their paws on the front of the cages, protrude their faces from between the bars, and appear to beg for the sucrose-coated tip of the syringe containing the drug (Marr et al., 1993).

It is important that illness is recognized early in laboratory rabbits. This can be crucial because pre-emptive treatment for diseases like coccidiosis is often contraindicated. As a prey species, rabbits will disguise any signs of illness if they can. A reduction of food intake may be an early sign. It is useful to weigh the rabbits whenever they are handled, for example when blood samples are being taken (Figure 8). This allows early detection of inappetence. In addition, small quantities of treats, such as carrots, lettuce or leafy hay, may be used to check if the rabbits are still eating (Figure 9). Normally all members of the group will gather round the treat. A rabbit who hangs back may not be feeling well and should be looked at a little more closely. Personnel who regularly distribute treats are recognized by the rabbits who will often gather at the front of the pens at the sound of the treats bag. This is an elegant way to check all members of the group, a task that should be done at least once every day. Technicians quickly learn to notice subtle changes in behavior and so become aware of health problems. Special work time should be set aside for them so that they can pet their charges every day, thereby fostering a positive human-animal relationship (Home Office, 1989). ***The gentle touch provided by the technicians is as important as the physical environment in giving the rabbits a sense of security*** in the presence of humans who, in other circumstances may subject them to uncomfortable, perhaps even painful procedures. Gentle, frequent handling of rabbits buffers their fear response during stressful situations (Anderson et al., 1972; Kertsen et al., 1989). Rabbits who receive special attention from personnel [frequent handling, petting, playing, gentle vocalization] show a markedly increased resistance to certain pathological processes than subjects who receive no extra attention (Nerem et al., 1980).

Figure 8. Regular, gentle health checks and weighing are important in monitoring the well-being of the rabbits and fostering a positive human-animal relationship.

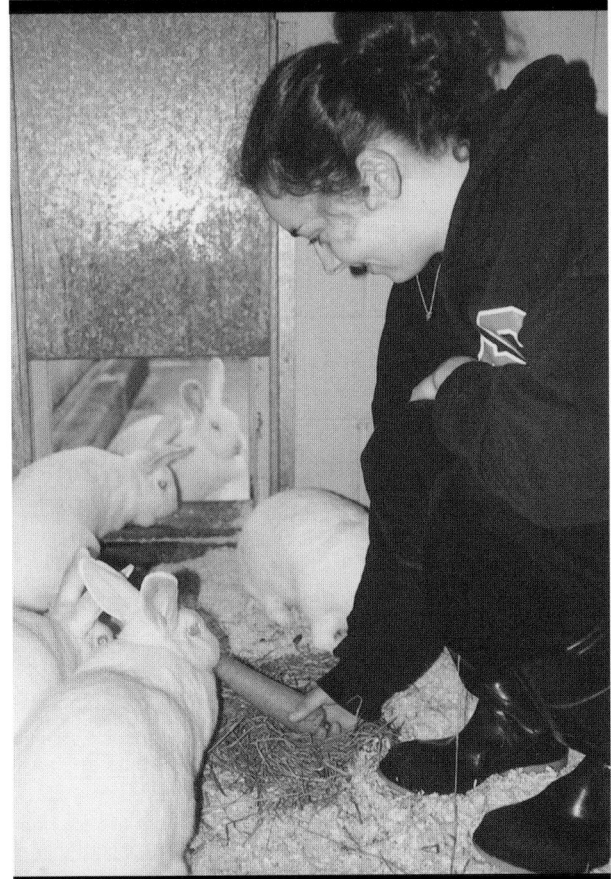

Figure 9. Providing treats helps win the confidence and trust of the rabbits and allows the technician to check their appetite.

References

Anderson CO, Denenberg VH, Zarrow MX 1972. Effects of handling and social isolation upon the rabbit's behaviour. Behaviour 43, 165-175

Batchelor GR 1999. The Laboratory Rabbit. In The UFAW Handbook on the Care and Management of Laboratory Animals Seventh Edition Poole T, English P (eds), 395-408. Blackwell Science, Oxford, UK

Bellhorn RW 1980. Lighting in the animal environment. Laboratory Animal Science 30, 440-450

Berthelsen H, Hansen LT 1999. The effect of hay on the behaviour of caged rabbits (Oryctolagus cuniculus). Animal Welfare 8, 149-157

Bigler L, Oester H 1994. Paarhaltung nicht reproduzierender Zibben im Käfig. Berliner und Münchner tierärztliche Wochenschrift 107, 202-205

Brooks DL, Huls W, Leamon C, Thomson J, Parker J, Twomey S 1993. Cage enrichment for female New Zealand White rabbits. Lab Animal 22(5), 30-38

Clough G 1982. Environmental effects on animals used in biomedical research. Biological Reviews 57, 487-523

Drescher B, Loeffler K 1991. Einfluß unterschiedlicher Haltungsverfahren und Bewegunsmöglichkeiten auf die Kompakta der Röhrenknochen von Versuchs- und Fleischkaninchen. Tierärztliche Umschau 46, 736-741

Gerson P 2000. The modification of "traditional" caging for experimental laboratory rabbits and assessment by behavioural study. Animal Technology 51, 13-36

Gunn D, Morton DB 1993. The behaviour of single-caged and group-housed laboratory rabbits. Proceedings of the Fifth Federation of European Laboratory Animal Science Association (FELASA) Symposium, 80-84

Gunn-Dore D 1997. Comfortable quarters for laboratory rabbits. In Comfortable Quarters for Laboratory Animals, Eighth Edition Reinhardt V (ed), 46-54. Animal Welfare Institute, Washington, DC
Full Text: http://www.awionline.org/pubs/cq/five.pdf

Gunn-Dore D 1999. Wire balls as enrichment for individually caged rabbits. Animal Technology 50, 162-163

Held SDE 1996. Group-Housing of Female Laboratory Rabbits—Studies on Behaviour and Immunocompetence. Ph.D. dissertation, University of Wales, Aberystwyth, UK

Held SDE, Turner RJ, Wootton RJ 2001. The behavioural repertoire of non-breeding group-housed female laboratory rabbits (Oryctolagus cuniculus). Animal Welfare 10, 437-443

Home Office 1989. Animals (Scientific Procedures) Act 1986. Code of Practice for the Housing and Care of Animals Used in Scientific Procedures. Her Majesty's Stationery Office, London, UK
Full Text: http://www.homeoffice.gov.uk/animact/hcasp.htm

Huls WL, Brooks DL, Bean-Knudsen D 1991. Response of adult New Zealand White rabbits to enrichment objects and pair-housing. Laboratory Animal Science 41, 609-612

Kertsen AMP, Meijsser FM, Metz JHM 1989. Effects of early handling on later open-field behaviour in rabbits. Applied Animal Behaviour Science 24, 157-167

Kraft R 1978. Vergleichende Verhaltensstudien an Wild- und Hauskaninchen I. Das Verhaltensinventar von Wild- und Hauskaninchen. Zeitschrift für Züchtungsbiologie 95, 140-162

Kraus AL, Weisbroth SH, Flatt RE 1994. Biology and disease of rabbits. In Laboratory Animal Medicine Fox JG, Cohen BJ, Loew FM (eds), 207-240. Academic Press, Orlando, FL

Krohn TC, Ritskes-Hoitinga J, Svendsen P 1999. The effect of feeding and housing on the behaviour of the laboratory rabbit. Laboratory Animals 33, 101-107

Lehmann M 1991. Social behaviour of young domestic rabbits under semi-natural conditions. Applied Animal Behavioural Science 32, 269-292

Lindfors L 1997. Behavioural effects of environmental enrichment for individually caged rabbits. Applied Animal Behaviour Science 52, 157-169

Loeffler K, Drescher B, Schulze G 1991. Einfluß unterschiedlicher Haltunsverfahren auf das Verhalten von Versuchs- und Fleischkaninchen. Tierärztliche Umschau 46, 471-478

Love JA, Hammond K 1991. Group-housing rabbits. Lab Animal 20(8), 37-43
Full Text: http://www.awionline.org/lab_animals/biblio/la20-8rab.html

Love JA 1994. Group Housing: Meeting the physical and social needs of the laboratory rabbit. Laboratory Animal Science 44, 5-11

Marr JM, Gnam EC, Calhoun J, Mader JT 1993. A non-stressful alternative to gastric gavage for oral administration of antibiotics in rabbits. Lab Animal 22(2), 47-49

Morton DB, Jennings M, Batchelor GR, Bell D, Birke L, Davies K, Eveleigh JR, Gunn D, Heath M, Howard B, Koder P, Phillips J, Poole T, Sainsbury AW, Sales GD, Smith DJA, Stauffacher M, Turner RJ 1993. Refinements in rabbit husbandry. Second report of the BVAAWF/FRAME/RSPCA/UFAW joint working group on refinement. Laboratory Animals 27, 301-329
Full Text: http://www.lal.org.uk/pdffiles/rabbit.PDF

Mykytowycz R 1958. Social behaviour of an experimental colony of wild rabbits (Oryctolagus cuniculus L.) 1. Establishment of the colony. CSIRO Wildlife Research 3, 7-25

Nerem RM, Levensque MJ, Cornhill JF 1980. Social environment as a factor of diet induced atherosclerosis. Science 208, 1475-1476

Podberscek AL, Blackshaw JK, Beattie AW 1991. The behaviour of group penned and individually caged laboratory rabbits. Applied Animal Behaviour Science 28, 353-363

Raje SS, Stewart KL 1997. Group housing for male New Zealand White rabbits. Lab Animal 26(4), 36-37
Full Text: http://www.awionline.org/lab_animals/biblio/la26-4rab.html

Rothfritz P, Loeffler K, Drescher B 1992. Einfluß unterschiedlicher Haltunsverfahren und Bewegungsmöglichkeiten auf die Spongiosastruktur der Rippen sowie Brust- und Lendenwirbel von Versuchs- und Fleischkaninchen. Tierärztliche Umschau 47, 758-768

Stauffacher M 1992. Group housing and enrichment cages for breeding, fattening and laboratory rabbits. Animal Welfare 1, 105-125

Stauffacher M 1993. Refinement bei der Haltung von Laborkaninchen. Ein Beitrag zur Umsetzung von Tierschutzforderungen in der Praxis. Der Tierschutzbeauftragte 2/3, 18-33

Stauffacher M, Bell DJ, Schulz K-DBV, Brain PF, Büttner D, Drescher B, Jilge B, Laurent J, Loeffler K, Militzer K, Morton DB, Nebendahl K, Schwartz K, Turner RJ, Völlm J 1994. Rabbits. In <u>The Accommodation of Laboratory Animals in Accordance with Animal Welfare Requirements. Proceedings of an International Workshop held at the Bundesgesundheitsamt, Berlin [The Berlin Workshop]</u> O'Donoghue PN (ed), 15-30. Bundesministerium für Ernährung, Landwirtschaft und Forsten, Bonn, Germany

Tamburrino PA, Michonski KJ, Cameron RA 1999. Adaptation of dog kennels for group housing of rabbits. <u>Abstracts of the AALAS [American Association for Laboratory Animal Science] Meeting</u>, 33

Turner RJ, Selby JI, Held SDE, Howells KJ, Eveleigh JR, Wootton RJ 1992. Preferred substrates for penned laboratory rabbits. <u>Animal Technology</u> 43, 185-192

Turner RJ, Held SDE, Hirst JE, Billinghurst G, Wootton RJ 1997. An immunological assessment of group-housed rabbits. <u>Laboratory Animals</u> 31, 362-372

United States Department of Agriculture 1991. Title 9, CFR (Code of Federal Register), Part 3. Animal Welfare; Standards; Final Rule. <u>Federal Register</u> 56(32), 6426-6505 **Full Text:** http://www.nal.usda.gov/awic/legislat/awadog.htm

Wemelsfelder F 1994. Animal boredom—A model of chronic suffering in captive animals and its consequences for environmental enrichment. <u>Humane Innovations and Alternatives in Animal Experimentation</u> 8, 587-591 **Full Text:** http://www.psyeta.org/hia/vol8/wemelsfelder.html

Wieser R 1986. <u>Funktionale Analyse des Verhalten als Grundlage zur Beurteilung der Tiergerechtheit. Eine Untersuchung zum Normalverhalten und Verhaltensstörungen bei Hauskaninchen-Zibben</u>. Doctoral Dissertation, Universität Bern, Bern, Switzerland

---

The authors of this chapter are animal caretakers, animal technicians and a veterinarian who have made a commitment over the years to refine the housing and handling conditions of rabbits to better address the animals' behavioral needs.

Comfortable Quarters for Cats in Research Institutions

Irene Rochlitz

Department of Clinical Veterinary Medicine, University of Cambridge, Madingley Road, Cambridge, CB3 OES, United Kingdom

The domestic cat has evolved from the African wild cat (*Felis silvestris libyca*), a semi-arboreal carnivore with an essentially solitary lifestyle. Cats do not have as wide a behavioural repertoire [facial expression, body posture, tail position] for visual communication as, for example, the highly social, group-living dog. They are more likely to respond to poor housing conditions by becoming inactive and by inhibiting normal behaviours such as feeding, grooming, exploring and playing than by showing abnormal behaviour (McCune, 1992; Rochlitz, 1997a). Sick cats will respond in a similar way. Therefore, keeping cats in a species-appropriate environment that encourages a wide range of normal behaviours will not only enhance the animals' welfare, making them better subjects for scientific investigation (Poole, 1997), but will also make it easier for caretakers to detect when an animal is unwell.

The emphasis in laboratory animal housing has been shifting from an "engineering" approach [specifying cage dimensions and features, and management procedures] to a "performance" approach [providing housing conditions and management procedures that enable the animals to reach certain performance standards] (National Research Council, 1996). The "performance" approach offers more flexibility, while the "engineering" approach provides a basis for the establishment of minimum requirements. This chapter presents recommendations for the housing and care of cats in laboratories, which are based on recent research and include both "engineering" and "performance" approaches.

Single-housing Versus Social-housing

Even though they have evolved from a solitary species, domestic cats are social animals who regularly interact with conspecifics (Leyhausen, 1979; Sandell, 1989). In most instances, cats will benefit from being housed with others provided there is sufficient space, easy access to feeding and elimination areas and an adequate number of retreats and rest places.

Many factors will determine the ideal group size, but it seems that 20 to 25 individuals is the maximal number for cats in laboratories (James, 1995; Hubrecht and Turner,

Figure 1. Cats who do not adapt to living harmoniously in groups or in pairs should be housed singly.

1998). Cats who fail to adapt satisfactorily to group-living generally adapt well to pair-housing. If a cat shows persistent incompatibility with other conspecifics she or he should be housed singly (Figure 1). If single-housing is necessary, the cage should be arranged in such a way that the cat has

Comfortable Quarters for Laboratory Animals Reinhardt V, Reinhardt A (eds), 50-55. Animal Welfare Institute, Washington, DC 20007

visual contact with other cats. While some authors suggest that tomcats should be housed singly, others have shown that they can be housed successfully with other males (Hart, 1980), with neutered males (Podberscek et al., 1991), or with neutered females. Queens in the last two weeks of pregnancy, and queens with unweaned kittens should be housed without other cats.

The introduction of a new animal to a group should be done slowly, under careful supervision. Initially, the resident cats should be able to get familiarized with the stranger without risk of overt aggression. To accomplish this, the new cat has to be kept in a separate cage within the group's enclosure. He or she must have access to a hiding box in the cage, in order to escape the attention of resident cats. Usually within two weeks, the newcomer can be safely released into the enclosure without risk. The use of synthetic analogues of naturally-occurring feline facial pheromones may facilitate the harmonious introduction of a strange cat into an established group (Pageat and Tessier, 1997).

Size of Primary Enclosure

The enclosure must be large enough to allow cats to express a wide range of normal postures and behaviours, such as stretching, exploring and playing, and to permit the caretaker to carry out routine procedures easily. In order to minimize social tension and prevent aggressive conflicts, and for reasons of hygiene, there should be a distance of at least 0.5 m between the various functional areas for feeding/drinking, resting, scratching, and defecating/urinating.

Group-living cats lack distinct dominance hierarchies and post-conflict mechanisms such as reconciliation (van den Bos and de Cock Buning, 1994a; van den Bos, 1998). They are not adapted to living in close proximity to one another and reduce the likelihood of aggression by keeping a distance between themselves (Figure 2; Leyhausen, 1979). Bernstein and Strack (1996) described the use of space and patterns of interaction of 14 unrelated cats who lived together in a single-storey house, at a density of one cat per 10 m^2. There was very little aggression and no fighting between the cats. Individuals peacefully co-existed because they were able to avoid one another for most of the time. If an enclosure is too small, there may be an increase in agonistic encounters, or the animals will attempt to avoid each other by decreasing their locomotor activity (Leyhausen 1979; van den Bos and de Cock Buning, 1994a). In a study of laboratory cats, daily activity levels dropped by 60 per cent when the animals were moved from a large enclosure [2.2 m^2 per cat] to a considerably smaller one [0.32 m^2 per cat] (Rochlitz, 1997b). Neutered, indoor-only pet cats kept in pairs maintain a distance of one to three meters from each other most of the time (Barry and Crowell-Davis, 1999). Kessler and Turner (1999) propose that there should be 1.7 m^2 per cat for group-housed cats in shelters. In research institutions, the minimum floor space should be 1.5 m^2 per weaned cat. It should be emphasised that the minimum floor space requirement for cats is determined by their socio-spatial needs rather than by body weight.

Cats use raised structures more often than the floor of their pens (Podberscek et al., 1991; Rochlitz et al., 1998).

Figure 2. There should be enough space for cats to have some privacy. Enclosures should contain structures that enable cats to use the vertical dimension.

High vantage points are used more frequently than low ones (Smith et al., 1994; Rochlitz, 1997b). As the vertical dimension is so important for cats, the enclosure of single-housed, pair- and group-housed animals should be at least 1.5 m high so that elevated resting surfaces can be installed well above ground level. Walk-in enclosures [2 m high] are ideal, as they also allow caretakers to easily enter and interact closely with the cats.

In some instances, it may be necessary to house a cat singly for recovery after a clinical procedure. This should be for as short a time as possible. The cage should still have at least 1.5 m^2 of floor space, but it does not have to have the height of a walk-in enclosure. However, it should be no less than 1 m high so that the cat can stretch fully in the vertical direction, and that a shelf can be installed in such a way that the cat can comfortably rest on it and freely move on and under it.

Cages should not be stacked one on top of the other. Placing the cages on a shelf at waist height or higher will make access easier for the caretaker.

Quality of Primary Enclosure

Beyond a certain minimum size, it is the quality rather than the quantity of space that is important for cats (Rochlitz, 2000). Domestic cats are agile, semi-arboreal animals who are skillful climbers. They use elevated areas as vantage points from which to monitor their surroundings and the approach of people and other animals (DeLuca and Kranda, 1992; Holmes, 1993; James, 1995). Enclosures should contain structures that enable cats to make maximal use of the vertical dimension, such as climbing frames, raised walkways, hammocks, and platforms or shelves placed at different heights (Figures 1 & 2). Slanting boards, steps and poles will help kittens and small cats to reach the raised areas.

There needs to be a sufficient number of rest areas for all cats in the enclosure, as cats like to rest alone rather than with others (Podberscek et al., 1991, Bernstein and Strack, 1996). Rest areas should have comfortable bedding (Figure 3). Cats who sleep on soft surfaces have longer periods of deep sleep than those who sleep on hard surfaces, suggesting that they feel more secure (Crouse et al., 1995). Cats prefer polyester fleece to cotton-looped towel, woven rush-matting, or corrugated cardboard as bedding material (Hawthorne et al., 1995). If there are not a sufficient number of comfortable rest places, the animals will use their litter trays for this purpose (DeLuca and Kranda, 1992; Rochlitz, 1997b). This is undesirable because it prompts other cats to urinate and defecate outside the litter trays.

Hiding is a behaviour that cats frequently display in response to changes in their environment, as well as to avoid interactions with other cats or with people (McCune, 1992; James, 1995; Rochlitz et al., 1998; Figure 4). A study of pair-housed cats found that the animals spent 50 per cent of the time out of sight of each other (Barry and Crowell-Davis, 1999). Visual barriers, such as vertical panels, trellises and curtains, can be used to divide the enclosure into separate spaces enabling cats to get out of sight of others. In a study of the behavioural and physiological correlates of stress in laboratory cats, hiding behaviour was negatively correlated with urinary cortisol concentration (Carlstead et al., 1993). This suggests that hiding functions as a stress-buffer. Therefore, cats should always have access to comfortable hiding places [e.g., boxes, deep-sided trays] in addition to open rest areas [e.g., shelves]. A raised, partially enclosed structure is useful as it conceals the cat who can monitor her or his surroundings at the same time.

Each cage of individually housed cats must have a litter tray. In group-housing, there should be at least one litter tray per two cats (Hoskins, 1996). Trays have to be cleaned once or several times daily, because some cats will not use a tray if it has already been soiled.

In a study of dominance in a group of female cats (van den Bos and de Cock Buning, 1994b), lower-ranking animals spent little time on the floor, were most often found on shelves, and appeared to be less mobile than higher-ranking individuals. Lower-ranking cats lost weight over time, and would use their rest sites for urination and defecation. Higher-ranking cats occupied the floor area, moved around the colony room more freely and tended to gain weight. These

Figure 3. Rest areas should have comfortable bedding.

Figure 4. Cats need retreat areas where they can hide.

Figure 5. Cats should have access to appropriate surfaces for scratching, as well as to a variety of toys.

Figure 6. Dexterity is required to retrieve pieces of food from the puzzle (photo by Geoffrey Loveridge, 1997).

findings indicate that it is advisable to place feeding stations, water bowls, resting shelves and litter trays in a number of different sites, to prevent certain cats from monopolising one resource and denying others access to it.

Surfaces for claw abrasion and olfactory marking [e.g., scratch posts, rush matting, carpet, wood] as well as a variety of play objects should be provided for social-housed cats (Figure 5) as well as for individually caged cats. Small, mobile objects with a complex surface texture are the most successful at promoting play (Hall and Bradshaw, 1998). De Monte and Le Pape (1997) found that a tennis ball was a more effective enrichment tool than a wooden log, for single-caged animals. Cats are intelligent animals and quickly lose interest in such play objects (De Monte and Le Pape, 1997). Varied toys should, therefore, be substituted regularly to ensure their novelty effect. Most cats play alone rather than in a group (Podberscek et al., 1991), so the cage has to be large enough for cats to play without disturbing others.

Another environmental enrichment technique is to increase the time animals spend in pseudo-predatory and feeding behaviour. McCune (1995) suggests putting pieces of dry food into containers with holes through which the cat must extract individual pieces; such a device stimulates play and pseudo-predatory behaviour (Figure 6).

Windows are an excellent way of providing interesting stimulation for confined cats (Loveridge, 1994). DeLuca and Kranda (1992) found that animals housed as a group spend most of the day sitting on a window perch, watching activity in the outside hallway. Windows should have deep sills or nearby elevated resting surfaces.

Consideration should be given to providing containers of grass. Most cats like to chew and ingest fresh grass, which helps them to eliminate furballs [trichobezoars]. Catnip (Nepeta cataria) is also enjoyed by many cats, either as a dried herb or when contained in toys.

The floor of all cat enclosures should be smooth, non-slip and easy to clean. Wire-mesh or grid floors are not suitable, as they are uncomfortable for cats, and may trap and injure their extremities [digits, paws and tails].

Control over the immediate environment is a feature of good animal housing (Broom and Johnson, 1993). An enclosure with separate functional areas [e.g., pen with an outdoor run; interconnected pens; vertical divisions within a pen], species-appropriate furniture [e.g., shelves; hiding box] and sensory access to the surroundings [e.g., window], allows cats some control over their physical and social environment. They can then make a variety of behavioural choices [e.g., climb on an elevated platform to get a better view; move behind a visual barrier to be alone] for optimal environmental adjustment and general well-being.

Contact with Humans

To facilitate handling and minimize stress during routine and experimental procedures, cats must be well socialised to humans. One of the best ways to achieve this is to ensure that young kittens are handled and spoken to on a daily basis. This is particularly important between two and seven weeks of age, the kittens' sensitive period for socialisation to people (Karsh and Turner, 1988; Figure 7). This regular, gentle handling should continue throughout the cat's life. Other factors that influence the sociability of cats to humans are summarized in Turner (2000).

Cats respond strongly to humans in their environment, and they organise their daily activity patterns around the caretaker's activity (Randall et al., 1990). In research facilities, cats demonstrate a clear preference for human contact over toys (DeLuca and Kranda, 1992). Hoskins (1995) examined the effect of human contact on the reactions of cats in a rescue shelter. Animals who were regularly handled by a familiar person allowed an unfamiliar person to hold them

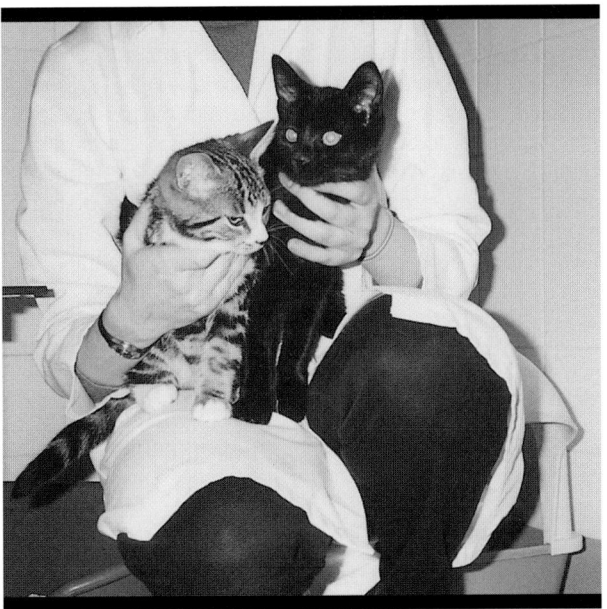

Figure 7. Young kittens should be stroked and handled gently on a daily basis.

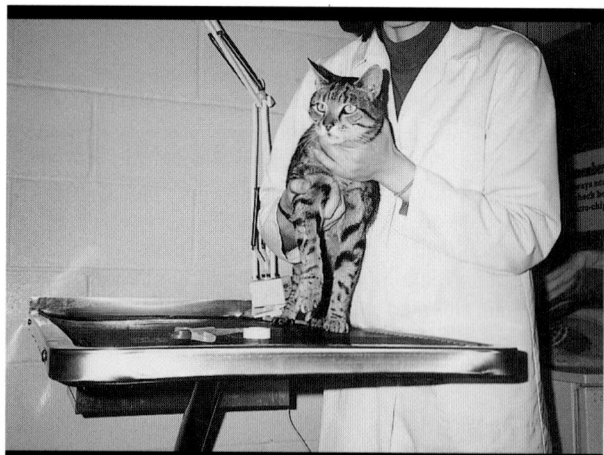

Figure 8. Cats who are well socialised to humans show little or no resistance during research-related handling procedures.

for a longer period than naïve cats. This effect is likely to be beneficial in the laboratory setting, where cats are handled on a regular basis by the [familiar] caretaker, and on an irregular basis by the [unfamiliar] researcher (Figure 8).

Cats show signs of stress [altered behaviour and raised urinary cortisol levels] when they are subjected to an unpredictable caretaking routine, and when technicians stop petting them and talking to them (Carlstead et al., 1993). Sudden changes in husbandry routine or the environment should be avoided. If changes need to be made, they should be introduced slowly and under careful supervision.

Caretakers need to be knowledgeable about the species-typical behaviours of cats. They should enjoy working with and for the animals. The importance of positive social interactions between care personnel and their animals cannot be over-stated. While cleaning and feeding times provide some opportunities for interactions, a period of time which is not part of routine procedures should be set aside every day for caretakers to be with their cats. Some cats may prefer to be petted and handled, while others may prefer to interact via a toy (Karsh and Turner, 1988). Most cats enjoy being groomed, and this can be a good time to detect whether an animal has lost weight or is otherwise unwell. Social contact with humans is particularly important for individually caged cats.

Source of Cats

Cats should be obtained only from designated breeding establishments [class A dealers].

Concluding Remarks

This chapter describes the needs of cats kept in research institutions for a stimulating, well-designed housing environment and appropriate considerate care from personnel. Attention to these housing and care requirements will result in healthy, friendly, well-adapted cats, thus enabling scientists to use fewer animals in their research and obtain better results.

Acknowledgements

I would like to thank the Blue Cross Adoption Centre, Cambridge, UK, for its help with photographs, and Professor D.M. Broom for providing the facilities to write this chapter.

References

Barry KJ, Crowell-Davis SL 1999. Gender differences in the social behaviour of the neutered indoor-only domestic cat. Applied Animal Behaviour Science 64, 193-211

Bernstein PL, Strack M 1996. A game of cat and house: spatial patterns and behaviour of 14 cats (Felis catus) in the home. Anthrozoos 9, 25-39

Broom DM, Johnson KG 1993. Stress And Animal Welfare. Chapman and Hall Ltd., London, UK

Carlstead K, Brown JL, Strawn W 1993. Behavioural and physiological correlates of stress in laboratory cats. Applied Animal Behaviour Science 38, 143-158

Crouse SJ, Atwill ER, Lagana M, Houpt KA 1995. Soft surfaces: A factor in feline psychological well-being. Contemporary Topics in Laboratory Animal Science 34(6), 94-97

De Monte M, Le Pape G 1997. Behavioural effects of cage enrichment in single-caged adult cats. Animal Welfare 6, 53-66

DeLuca AM, Kranda KC 1992. Environmental enrichment in a large animal facility. Laboratory Animal 21, 38-44

Hall SL, Bradshaw JWS 1998. The influence of hunger on object play by adult domestic cats. Applied Animal Behaviour Science 58, 143-150

Hart BL 1980. Feline Behaviour. A Practitioner Monograph. Veterinary Practice Publishing, Santa Barbara, CA

Hawthorne AJ, Loveridge GG, Horrocks LJ 1995. The behaviour of domestic cats in response to a variety of surface-textures. In Proceedings of the Second International Conference on Environmental Enrichment Holst B (ed), 84-94. Copenhagen Zoo, Copenhagen, Denmark

Holmes RJ 1993. Environmental enrichment for confined dogs and cats. In Animal Behaviour—The TG Hungerford Refresher Course for Veterinarians, Proceedings 214 Holmes RJ (ed), 191-197. Post Graduate Committee in Veterinary Science, Sydney, Australia

Hoskins CM 1995. The Effects of Positive Handling on the Behaviour of Domestic Cats in Rescue Centres. MSc. thesis, University of Edinburgh, UK

Hoskins JD 1996. Population medicine and infectious diseases. Journal of the American Veterinary Medical Association 208, 510-512

Hubrecht RC, Turner DC 1998. Companion animal welfare in private and institutional settings. In Companion Animals in Human Health Wilson CC, Turner DC (eds), 267-289. Sage Publications, Thousand Oaks, CA

James AE 1995. The laboratory cat. The Australian and New Zealand Council for the Care of Animals in Research and Teaching News 8, 1-8

Karsh EB, Turner DC 1988. The human-cat relationship. In The Domestic Cat: The Biology of its Behaviour, First Edition. Turner DC, Bateson P (eds), 159-177. Cambridge University Press, Cambridge, UK

Kessler MR, Turner DC 1997. Stress and adaptation of cats (*Felis silvestris catus*) housed singly, in pairs and in groups in boarding catteries. Animal Welfare 6, 243-254.

Kessler MR, Turner DC 1999. Effects of density and cage size on stress in domestic cats (*Felis silvestris catus*) housed in animal shelters and boarding catteries. Animal Welfare 8, 259-267

Leyhausen P 1979. Cat Behaviour: The Predatory and Social Behaviour of Domestic and Wild Cats. Garland STPM Press, New York, NY

Loveridge GG 1994. Provision of environmentally enriched housing for cats. Animal Technology 45, 69-87

McCune S 1992. Temperament and the Welfare of Caged Cats. Ph.D. thesis, University of Cambridge, Cambridge, UK

McCune S 1995. Enriching the environment of the laboratory cat. AWIC Resource Series No. 2—Environmental Enrichment Information Resources for Laboratory Animals 1965-1995, Birds, Cats, Dogs, Farm Animals, Ferrets, Rabbits, and Rodents 2, 27-33
FT: http://www.nal.usda.gov/awic/pubs/enrich/labcat.htm

National Research Council 1996. Guide for the Care and Use of Laboratory Animals. National Academy Press, Washington, DC
Full Text: http://www.nap.edu/readingroom/books/labrats/

Pageat P, Tessier Y 1997. Usefulness of the F4 synthetic pheromone for prevention of intraspecific aggression in poorly socialised cats. Proceedings of the 1st International Conference on Veterinary Behavioural Medicine, 64-72

Podberscek AL, Blackshaw JK, Beattie AW 1991. The behaviour of laboratory colony cats and their reactions to a familiar and unfamiliar person. Applied Animal Behaviour Science 31, 119-130

Poole TB 1997. Happy animals make good science. Laboratory Animals 31, 116-124

Randall WR, Cunningham JT, Randall S 1990. Sounds from an animal colony entrain a circadian rhythm in the cat, *Felis catus* L. Journal of Interdisciplinary Cycle Research 21, 55-64

Rochlitz I 1997a. The Welfare of Cats Kept in Confined Environments. Ph.D. thesis, University of Cambridge, Cambridge, UK

Rochlitz I 1997b. The welfare of cats in two research laboratories. BSAVA [British Small Animal Veterinary Association] Congress Proceedings, 309

Rochlitz I, Podberscek AL, Broom DM 1998. The welfare of cats in a quarantine cattery. The Veterinary Record 143, 35-39

Rochlitz I 2000. Recommendations for the housing and care of domestic cats in laboratories. Laboratory Animals 34, 1-9

Sandell M 1989. The mating tactics and spacing patterns of solitary carnivores. In Carnivore Behaviour, Ecology and Evolution Gittelman JL (ed), 164-182. Chapman & Hall, London, UK

Smith DFE, Durman KJ, Roy DB, Bradshaw JWS 1994. Behavioural aspects of the welfare of rescued cats. The Journal of the Feline Advisory Bureau 31, 25-28

Turner DC 2000. The human-cat relationship. In The Domestic Cat: The Biology of its Behaviour, Second Edition. Turner DC, Bateson P (eds), 193-206. Cambridge University Press, Cambridge, UK

van den Bos R, de Cock Buning T 1994a. Social and non-social behaviour of domestic cats (*Felis catus* L.): a review of the literature and experimental findings. In Welfare and Science-Proceedings of the Fifth FELASA Symposium Bunyan J (ed), 53-57. Royal Society of Medicine Press Ltd, London, UK

van den Bos R, de Cock Buning T 1994b. Social behaviour of domestic cats (*Felis lybica* f. *catus* L.): a study of dominance in a group of female laboratory cats. Ethology 98, 14-37

van den Bos R 1998. Post-conflict stress-response in confined group-living cats (*Felis silvestris catus*) Applied Animal Behaviour Science 59, 323-330

Irene Rochlitz is a veterinarian with an interest in animal behaviour and welfare. She is a research associate with the Animal Welfare and Human-Animal Interactions Group at the University of Cambridge, UK. She combines research into issues affecting the welfare of cats with clinical work in veterinary practice.

Comfortable Quarters for Dogs in Research Institutions

Robert Hubrecht

Universities Federation for Animal Welfare, The Old School, Brewhouse Hill, Wheathampstead, Herts, AL4 8AN, United Kingdom

The dog is one of the oldest domesticated animals and has probably been associated with man for at least 14,000 years. For many people their pet dog has a special status and is often considered as a member of the family. In consequence, the dog when used as an experimental animal receives special protection in some countries. For example, in the UK, special justification has to be provided before dogs can be assigned to a study involving pain, suffering or distress. Dogs used in studies in Europe have to originate from designated or registered supplying establishments, which in the UK are inspected by the Home Office.

A good day-to-day care person should have a sound understanding of the dogs' biology (Serpell, 1995), particularly of their social nature (Boitani et al., 1995; Macdonald and Carr, 1995). Moreover, experience and research show that both conspecific and human social contact are extremely important for the well-being of dogs. Dogs are inquisitive animals who actively seek information about their surroundings and so will react badly to barren or sensory restricted environments. They use a variety of modes of communication. Olfaction, one of the most important of these, is unfortunately one which we have the most difficulty in empathising with. Husbandry staff should be experienced enough to be able to identify the meaning of the various dog vocalisations, such as distress and threat vocalisations, the high pitched bark given by a dog who is separated from his or her social companions, and unspecific vocalisation in response to arousal. Staff should be able to identify visual signals, which include posture and facial expressions (Fox and Bekoff, 1975). For example, dogs who adopt a low posture are likely to be unsure of themselves and may require extra reassurance or training.

The attending care person is probably better at his/her job if he/she has experience with pet dogs outside the laboratory. This can provide valuable insights into the dogs' needs, variations in character, and a better understanding and rapport with the animals in her or his care.

Why Good Dog Accommodation is Good for Welfare as Well as Science

Canine enclosures have often been designed primarily for easy husbandry and to maintain the animal's physical health. These considerations are of course essential, but the dog is the prime user of the enclosure, and spends much longer periods in that environment than the staff who service it. It is vital, therefore, that designers consider the dog's normal behavior and are aware of the extent to which the enclosure might restrict such behavior. This is important not just from the point of view of animal welfare. It is well established that housing environments that do not meet the social or physical needs of an animal can lead to changes in physiology and behavior, thereby influencing research data. Dogs who are not able to cope with poorly designed laboratory housing are prone to develop stereotypical locomotory behaviors such as pacing, circling or wall bouncing (Hubrecht et al., 1992; Hubrecht, 1995a). In one case observed by this author, these behaviors were so extreme that the dog used three times his normal daily food ration to fund the metabolic expenditure resulting from the stereotypy. Clearly such abnormal behavior would have consequences for certain scientific outcomes. While extreme cases such as this are sometimes easily detected from patterns in the sawdust (Figure 1), others are often missed because the dog ceases to perform the stereotypy when care staff enter the kennel. To help identify stereotyped behavior, it can be useful to employ closed circuit television or video techniques to monitor the dogs from another room. Running a videotape in fast forward will make it easier to detect abnormally repetitive behavior patterns.

It must be remembered that dogs have different personalities that are a product of their genetic make-up and personal experiences. There will be an inevitable variation in the dogs' responses to kennelling conditions which in turn may influence their responses to experimentation and testing. Ideally, the environment should be one in which the most nervous dog in the colony can live without being unduly stressed.

Comfortable Quarters for Laboratory Animals Reinhardt V, Reinhardt A (eds), 56-64. Animal Welfare Institute, Washington, DC 20007

Figure 1. The pattern in the sawdust provides clear evidence that this dog has a stereotyped locomotor pattern. This is a cause for concern, and the housing design and management practices should be re-addressed. Stereotypies may not always be the product of the current housing, as once established they can be hard to eradicate.

purposes should ultimately become unnecessary except where there is a specific justification on scientific grounds [e.g., drugs that have an effect on aggressiveness or that are expected to cause vomiting].

Aggression can cause serious complications when dogs are housed in groups. It is critical to ensure that there is an adequate husbandry routine to monitor the animals and forestall potential problems. A video and sound monitoring system can be very helpful.

There are no clear data indicating what an optimum group size might be. Pair-housing seems to be a reasonable compromise, as dogs in pairs spend a similar proportion of their time interacting with each other as dogs kept in groups of 5-11 animals (Hubrecht, 1993). For this reason, the minimum dimension of all primary enclosures should be sufficient to house two dogs together in a space that allows them to express most of their normal behavioral repertoire and permits adequate enrichment (Figure 2). Pens should be designed so that management practices can be flexible. For example, groups of pens within a room should be arranged in such a way that it is possible to move an animal to another pen temporarily during wet cleaning, and so avoid exposing the subject to buckets of water, high pressure hoses or other aversive stimuli. It should also be possible to open doors or pop holes between pens so that larger "super-pens" can be created as desired, either to give the animals more space and increase its complexity or to allow groups of dogs to run together (Figures 2 & 3).

Social Housing as the Basis for Kennel Design

Primary enclosures should always be large and flexible enough to house dogs socially in harmonious groups. Single-housing for prolonged periods is apt to be deleterious to the dog and is associated with an increased incidence of behavioral abnormalities (Hetts et al., 1992; Hubrecht, 1995a). Single-housing may sometimes be necessary for clinical, behavioral or scientific reasons, but the duration should always be kept to a minimum, and the reasons for single-housing should always be challenged. For example, it used to be accepted that dogs used in toxicology were housed individually, but contract research organisations in the UK have for many years now routinely housed their dogs two or more to a pen (Hubrecht, 1995b). "Dogs on many GLP [good laboratory practice] toxicological studies can be housed in groups. This concept is, at the very least, worthy of consideration. In fact, we highly recommend it for most longer term GLP toxicological studies" (Hickey 1993, p. 77). While it is still common to separate the dogs for feeding and dosing, the period has been progressively shortened [currently 4 hours or less]. With appropriate training and management, the separation of dogs for toxicology

Space Considerations

Provision of adequate space is essential for dogs as it affects not only their behavior, but also determines whether the animals can be housed in social groups and whether there

Figure 2. Housing accommodation at Novo Nordisk, Denmark, designed to meet the behavioral needs of dogs. Note that the pens are linked in pairs with access to an external run through the pop-hole at the back of the pen. Platforms offer visibility across the room. Suspended chews and other toys are provided as means of environmental enrichment. A dog bed with cushion offers a comfortable resting place.

Figure 3. The platform within the exercise area enables the dog to monitor events within the room, and provides him/her with more choice of location and height. The platform is shielded thus also providing a refuge. The pop hole [right side of the photograph] allows access to the sleeping area with a lower platform. It also makes it possible for adjacent pens to be run together. Note the suspended enrichment device. Sprung chains maintain the chew/toy a few inches off the ground but enable the dog to use a paw to keep it temporarily on the ground whilst chewing.

is sufficient room for enrichment devices. Confinement intrinsically restricts the dog's ability to perform species-typical behaviors and to adjust social contact with other dogs (Bebak and Beck, 1993). Small or shallow-depth pens may also not allow the dog to retreat from events that he or she considers alarming at the front of the pen. Cramped enclosures are associated with a higher prevalence of circling and other stereotypies than relatively large enclosures (Hubrecht et al., 1992). This indicates that too small living areas affect the dogs' behavioral health and hence their general well-being. If dogs show stereotyped behavior, then there is good reason to re-examine the particular type of housing and attempt to improve it (Figure 1).

Minimum dimensions for primary enclosures are recommended as stipulated by the Home Office (1989; Table 1). The allocated floor areas provide a reasonable compromise between financial investment and provision of sufficient space for locomotion, socialisation and enrichment purposes. It is arguable whether pen size should be based on such a simple variable as body weight. For example, young animals are likely to require more space than old animals as they are more active and need extra space for play. The space recommendations listed in Table 1 are based on professional experience. They have the benefit of encouraging pair-housing, because the minimum sizes for single cages provide enough room for two adult dogs. For example, the minimum floor area for adult beagles of the weight category 10-25 kg is 4.5 m^2, both for a singly-housed dog and for two pair-housed dogs. The addition of a third dog, however, necessitates that the floor area be increased by 2.25 m^2, i.e., to 6.75 m^2. The area of minimum-size pens should never be reduced, even temporarily through the use of partitions [e.g., for toxicology dosing of pair-housed subjects], as it then becomes extremely difficult to provide a structured and enriched environment. The preferred option would then be to double the size of the enclosure so that each dog has access to the minimum space at all times.

The height of the enclosure should at least allow the dog(s) to stand on hind legs without touching the roof (Table 1).

Exercise

One obvious effect of confinement is to restrict locomotor behavior. Small enclosures not only discourage exercise

BODY WEIGHT KG (LB.) OF DOG	MINIMUM FLOOR AREA M^2 (FT.^{2}) PER DOG		MINIMUM HEIGHT CM (IN.)
	HOUSED SINGLY	HOUSED IN GROUPS*	
less than 5 (11)	4.5 (48.4)	1.0 (10.8)	150 (59.1)
5-10 (11-22)	4.5 (48.4)	1.9 (20.5)	150 (59.1)
10-25 (22-45)	4.5 (48.4)	2.25 (24.2)	200 (78.7)
25-35 (45-77)	6.5 (70.0)	3.25 (35.0)	200 (78.7)
more than 35 (77)	8.0 (86.1)	4.0 (43.1)	200 (78.7)

* Floor area must be not less than that specified for a singly housed dog.

Table 1. Minimum space recommendations for laboratory dogs.

Figure 4. Walking dogs on a leash, while perhaps not possible in all institutions, provides enrichment for the animals, offers exercise, improves staff morale and helps with re-homing.

because there is no possibility of travelling to another location, but they also restrict the type of locomotion that is possible and the ability of the dog to control his or her social interactions. Increasing pen dimensions beyond the minimum standards acceptable in the U.S. (United States Department of Agriculture, 1991) does not seem to make much difference in terms of the dog's physical fitness (Clark et al., 1991), aggression, or play (Bebak and Beck, 1993). Nonetheless, an ethological study of mixed breed dogs housed in pens with spacious, outdoor runs [744 m^2] has shown that both the activity of the animals and their range of species-typical locomotory behaviors was greater than that shown by dogs in small standard pens of less than 7 m^2 (Hubrecht et al., 1992).

It is reasonable to assume that space *per se* does not stimulate a dog to run around and exercise, but that the presence of structures or the presence of other dogs or care personnel will entice the dog to explore the available space and make active use of it. Any exercise program will have to take these basic ethological principles into account. It is highly recommended that laboratory dogs be exercised on a regular basis in a well-structured environment with other dogs and/or with friendly care personnel (Figures 4-8; United States Department of Agriculture, 1991; Loveridge, 1994,1998; Trussell et al., 1999; University of Florida, 2000).

Structures and Enrichment within the Dog Enclosure

Sufficiently large enclosures offer possibilities for the provision of structures and enrichment devices (Figures 2,3,5,6,7). They also allow the provision of separate sleeping and exercise areas, so that dog(s) can defecate and urinate away from the sleeping area. This makes the environment more complex and interesting and gives the animal(s) some choice and the option of exercise. Most importantly, a large enclosure permits social housing.

Objections are sometimes made to enrichment items such as chews/toys because it is considered that they can trigger aggression and cause hygiene problems, or simply because the dogs lose interest in them. Appropriate presentation of enrichment items can address these objections. For example, the toys can be suspended from the ceiling by sprung chains (Figure 3), which prevents soiling and makes it impossible for one dog to monopolise them (Hubrecht, 1993). The toys should be suspended just off the floor of the enclosure so that the dog can hold and gnaw the chew while lying down in a species-specific fashion. Dogs are very motivated by food. Toys or chews that have an appetising aroma or taste, therefore, receive a lot of attention. Such items can reduce the time during which the animals are inactive and decrease destructive behavior of cage fixtures and furnishings. The benefits of enrichment of beagles' enclosures have been demonstrated (Hubrecht, 1993,1995). Toys or chews [rawhide, plastic pipe and Gumabone tugtoy] were used by puppies for 64% of their time and by sub-adult dogs [7-13 months old] 24% of their time. No habituation effect was noticed over a two-month period.

Any enrichment device has to be practical, provide a measurable benefit for the animal and should not interfere with the aims of the research protocol.

Dogs are inquisitive animals who show a keen interest in their surroundings. Accordingly kennels should not restrict the animals' ability to obtain important information about their surroundings. High walls or solid partitions between pens result in the dogs being unable to see to the end of their rooms. This can cause them to spend a relatively high proportion of their time on hind legs or in apparently repetitive, possibly stereotypical jumping behavior. Obvious ways around this problem include reducing the height of partitions between pens for at least a part of their length and providing platforms (Figures 2 & 3). Hubrecht (1993) has shown that platforms are extensively used [55% of the time] by laboratory dogs to play and rest on, and that they do not pose any risk, even for dogs with gastric fistulas. Platforms make the third dimension accessible and increase spatial complexity, thus allowing the dogs more choices within their environment. If properly installed so that they do not block the existing floor area, platforms also provide additional floor area (Figures 2 & 3).

Dog housing should always be designed so that the occupants can retreat to an area that gives them a sense of security. This need not cause a problem of visibility for the day-to-day care person, as it can simply be an area with a few barriers shielding the animal from view on two or three sides (Figure 3). It is particularly important to offer such structures when dogs are housed in groups so that the animals can better control their social interactions.

Indoor-outdoor pens are an option in some cases and it is generally believed that offering dogs access to the outside is an enriching experience for them (Figures 5-7). Objections on the grounds of noise nuisance to neighbouring residents can be countered by appropriate design of the kennel structures or through the use of noise reflecting barriers such as banks of earth.

Figures 5-7. An outdoor exercise area at Novo Nordisk, Denmark, showing the large space [2000 m^2] provided for the dogs, with structures and enclosures ["pig huts" with flat roofs and chutes, hills with drain pipes and trees] within the pen. Dogs have access to this area 5 times a week for a minimum of 1-2 hours daily.

Flooring

The choice between solid or grid floors was considered at the Berlin Workshop (Gärtner et al., 1994). Open floored systems are sometimes preferred because they are cheaper to maintain and clean, but the majority of the experts recommended solid or at least only partly gridded floors and agreed that dogs prefer solid flooring. Whatever the flooring type, a safe, solid area of sufficient size for all dogs to comfortably and simultaneously lie down should be provided. When solid floors are used, a substrate such as sawdust helps to soak up urine and some of the moisture in faeces. The sawdust is generally not used in sufficient quantities to provide bedding for the dog and is generally not required for this purpose. However, comfortable bedding is recommended, especially for puppies, sick animals and old animals (Figure 2; Loveridge, 1994; Eisele, 2001).

Wet cleaning of enclosures should not be necessary on a daily basis. Many establishments now carry it out weekly or at longer intervals, using regular dry cleaning to remove soiled patches of sawdust.

Social Interactions of Dogs with Animal Care Staff

Considering the strong social disposition of dogs, it is not surprising that socially restricted rearing leads to the development of behavioral abnormalities and crippled behavioral repertoires (Thompson et al., 1956; Fuller, 1967). Similarly, if dogs are not provided with an adequate early experience of humans they will later be fearful of people and as a result can be difficult to handle (Freedman et al., 1961; Scott and Fuller, 1965; Wolfle, 1987). Puppies go through a stage between the 3rd and 12th week of life—the so-called primary socialisation period—when it is particularly easy for them to develop relationships with other individuals (Scott and Fuller, 1965; Wright, 1980). At the same time, the puppy becomes attached to the familiar home area. While there is debate as to whether this is really a "critical period" it does seem to be a time of special importance in the puppy's development (Markwell and Thorne, 1987).

Although the general time of this socialisation period is agreed upon, there is surprisingly little known about how much human-contact time is needed to adequately socialise a dog with humans. Some studies suggest that socialisation with humans can be achieved through relatively small amounts of time: 40 minutes or less per week spent with a litter (Scott and Fuller, 1965), or 5 minutes per week spent with each puppy (Wolfle, 1990). Hubrecht (1995) worked with male beagle puppies who had been considered by a pharmaceutical firm to have already received adequate socialisation. Even so, an extra 2 minutes spent in the pen each weekday combined with 30 seconds of petting each puppy [i.e., 2.5 minutes intensive contact with the human handler per puppy per week] resulted in behavioral changes 6-11 months later. These changes could be interpreted as intensified seeking of human contact.

Dogs who have been socialised to humans while puppies readily socialise with them as adults. It is generally

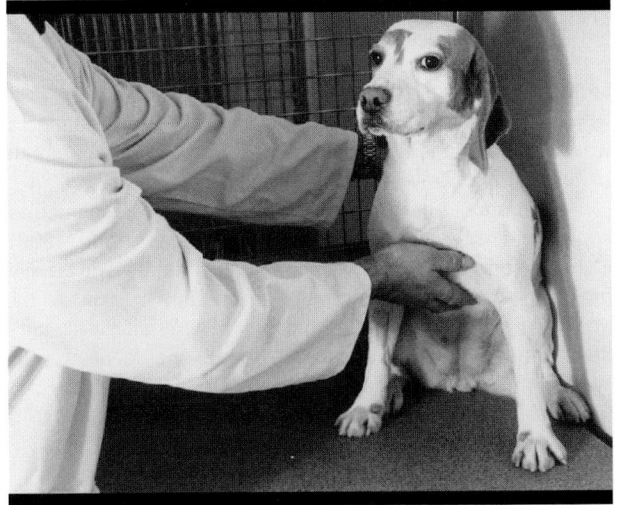
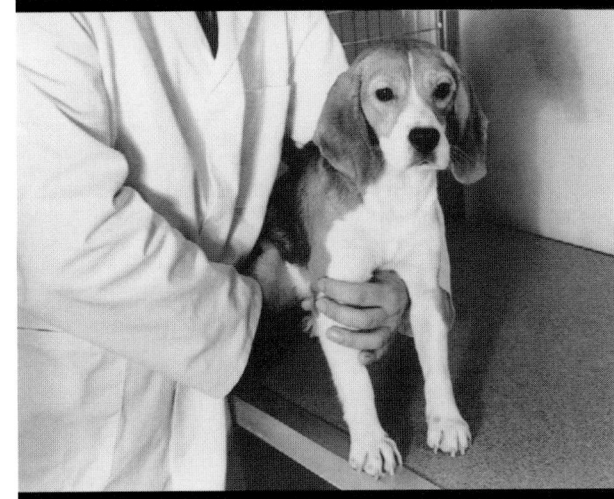
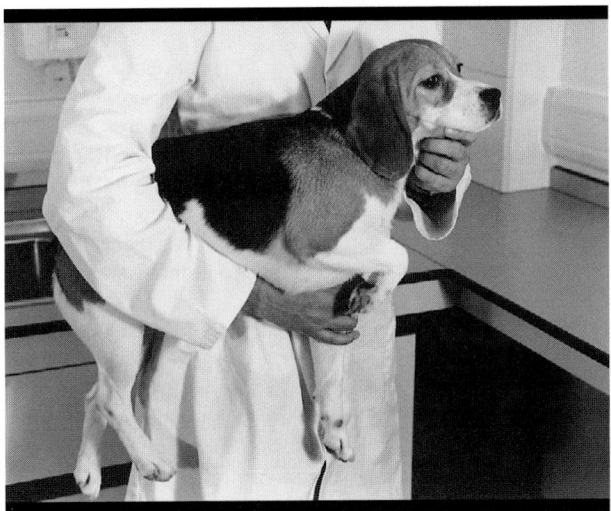
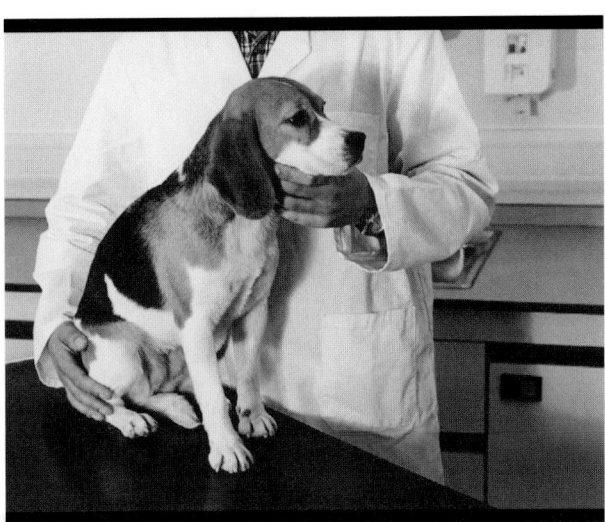

Figures 8-11. Professional handling techniques of laboratory dogs should be based on a positive, trustful human-animal relationship. The dog should feel at ease when being approached (8), picked up (9), carried (10), and restrained (11).

agreed by professionals that human socialisation with adult dogs improves handleability and provides an important form of enrichment (Fox, 1986; Loveridge, 1998). However, in kennels housing large numbers of dogs, the pressures on staff are often such that contact with the dogs becomes very limited (Hubrecht et al., 1992). Managers should be aware that this can be a serious problem and should implement a socialisation program.

Minimisation of Stress During Interactions Between People and Dogs

The easiest way to minimise stress in the dog is to ensure that he or she reacts well to handling. All dogs should, therefore, experience adequate socialisation with humans and other dogs during the "primary socialisation period" [approximately 3-12 weeks] and receive regular gentle and sympathetic handling thereafter. The same member of staff should always carry out the handling of an animal or, if this is not possible at times, by another person who applies the same handling techniques.

Training dogs so that they become used to experimental and clinical procedures is essential in order to avoid stress (Figures 8-14). If, for example, a procedure involves temporary restraint—which is known to be potentially stressful (Knol, 1989)—the familiar handler should first gently introduce the dog to this situation. The handler should always remain with the animal during the procedure. If this is not possible, the person carrying out the procedure should also pick up the dog from the pen and return the dog afterwards in the same manner as the regular handler would normally do. It is now becoming accepted that there should be close liaison between the breeders and users of laboratory dogs to ensure that handling methods are standardised. Young dogs should become used to typical procedures such as weighing or the taking of temperature. Sometimes it might be beneficial to accustom puppies to

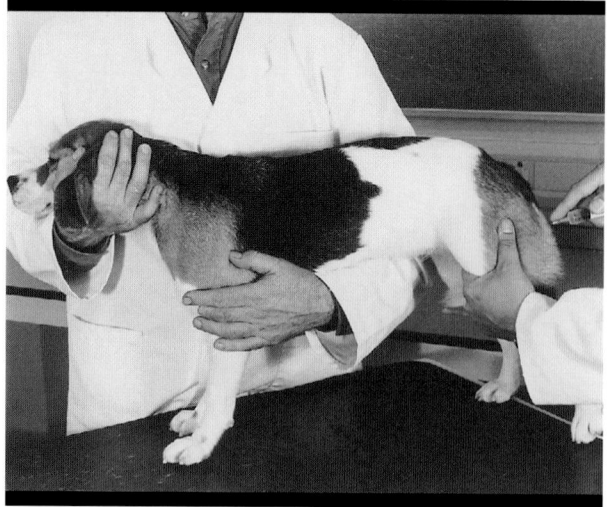

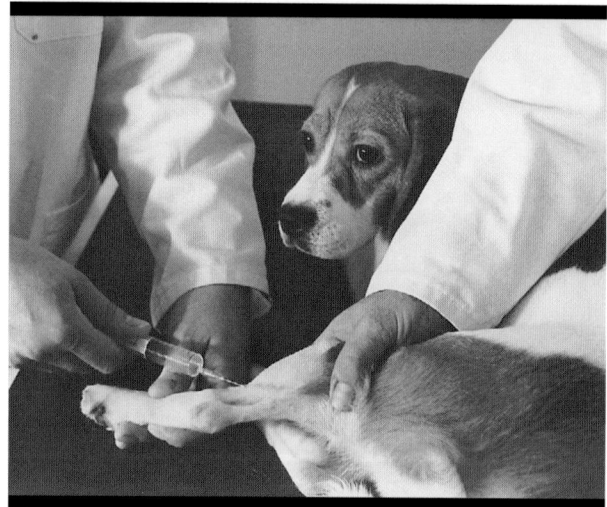

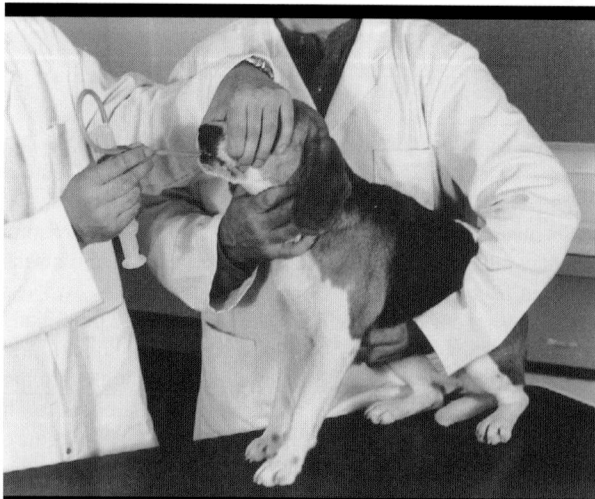

Figures 12-14. Laboratory dogs should be trained through positive reinforcement to accept routine procedures such as intramuscular injection (12), intravenous injection (13) and oral dosing (14) without showing noticeable stress reactions.

more sophisticated procedures. In these cases a cost/benefit decision should be taken as to whether any short-term welfare costs to the puppy are balanced by welfare benefits later in life. Socialised dogs can be more difficult to work with as they tend to be more boisterous. It may be beneficial when training dogs to provide clear cues so that the animals can distinguish between "work" and "play" times.

Staff should be proficient in basic handling techniques, such as those shown in Figures 8-14 (MacArthur, 1999). Care personnel should approach the dog steadily and quietly, make confident contact and should frequently reassure the dog by petting and talking quietly to him/her. The staff's demeanour while in the animal rooms should always be calm, confident and quiet. The aim is to establish a bond of trust with the dogs so that handling is a positive rather than a negative experience for all parties involved.

Noise in Kennels

The noise in kennels resulting from barking can be a nuisance and is potentially damaging to human hearing (van der Heiden, 1992). Dogs often bark at each other through adjacent pens or at people passing by. Barking is also associated with feeding times and is probably a result of the generally high levels of excitement. Very often barking spreads to other dogs, and in some animals the act of barking may itself function as a self-stimulus to further barking (Scott and Fuller, 1965).

Dogs can detect sounds ranging in frequency from 0.04 kHz up to around 50 kHz, which is well beyond the upper frequency limit of human hearing. They can hear sounds that are up to four times quieter than the human ear can detect (Fay, 1988). In dog kennels, sound levels within the human hearing range can regularly reach values between 85 and 122 decibels (Ottewill, 1968; Peterson, 1980; Sales et al., 1997). Most of the high level noise is probably produced by the dogs themselves, although other events such as cleaning, high pressure hose, doors banging and pagers may also contribute substantially to the acoustic environment (Sales et al., 1988). The noise in dog kennels can often be at a level that has the potential of causing damage and stress not only to humans but also to other animals who have less acute auditory sensitivity than dogs (Gamble, 1982; Milligan et al., 1993). It is also known that noise stress in humans can lead to physiological and health problems (van Raaij and Oortgiesen, 1996). It seems probable that dogs, like humans, are at risk of noise stress and possible hearing damage. Sound control should, therefore, be a priority when designing dog kennels. Noise can be limited by the use of sound-absorbent materials that must permit cleaning and should either be out of the dogs' reach or resistant to destruction. Efforts should also be made to reduce transmission of sound by the use of acoustic doors and the use of cavity walls. Corridors can channel the sound from one area to another, but as sound travels less easily around corners, the use of straight corridors should be avoided. Major noise-producing equipment should be sited as far away from the animals as possible.

Conclusions

Dog kennel design should:

- keep the animal(s) in good physical and mental health,
- permit easy handling of the dog(s) by personnel,
- be large enough to allow group housing of compatible dogs,
- be flexible enough to allow pens to be joined together to create larger runs,
- permit choice of location and provide interest within the enclosure,
- provide some refuge from kennel mates through the use of visual barriers,
- allow the dog good visibility of the room and of the area outside the pen,
- reduce to a minimum sound egress and ingress.

Dog Supply and Re-homing of "Used" Dogs

In Europe there is a requirement that dogs provided for scientific use are bred in, and obtained from, a designated or registered breeding establishment (Editors' note: These correspond to class A dealers in the U.S.). There is much to be said for this provision as:

1. The dogs from registered breeders are likely to be of a good and standard quality and free of disease.
2. They will not have experienced anything other than an institutional life and, therefore, will not suffer due to separation from a family life or because of loss of freedom.
3. The standards of the designated breeding establishment can be maintained through a process of licensing and inspection.
4. The public has reasonable assurance that their pet will not end up in research establishments.

There can be problems in retiring laboratory dogs to homes. Any decision to do so should take into account the fact that there may be welfare problems to the dog. Dogs who have spent a substantial portion of their life in institutions often show a number of behavioral disorders when being re-homed. These may include difficulties with house-training, separation-anxiety problems, and aggression triggered by fear-inducing unfamiliar experiences. On the other hand, there are some establishments that have successfully introduced re-homing programs, and others are currently examining the feasibility of such a scheme. In order to re-home research dogs safely and humanely, the institution must develop and maintain a comprehensive socialization and training program. As with any re-homing program, the proper matching of the retired dog with the new owner is the basic condition for success.

References

Bebak J, Beck AM 1993. The effect of cage size on play and aggression between dogs in purpose-bred beagles. Laboratory Animal Science 43, 457-459

Boitani I, Francisci F, Ciucci P, Anreoli G 1995. Population biology and ecology of feral dogs in central Italy. In The Domestic Dog: Evolution, Behaviour, and Interactions with People Serpell J (ed), 217-244. Cambridge University Press, Cambridge, UK

Clark JD, Calpi JP, Armstrong RB 1991. Influence of type of enclosure on exercise fitness of dogs. American Journal of Veterinary Research 52, 1024-1028

Eisele PH 2001. A practical dog bed for environmental enrichment for geriatric beagles, with applications for puppies and other small dogs. Contemporary Topics in Laboratory Animal Science 40(3), 36-38

Fay R 1988. Hearing in Vertebrates: A Psychophysics Data Book Hill-Fay Associates, Chicago, IL

Fox MW 1986. Laboratory Animal Husbandry Ethology, Welfare and Experimental Variables State University of New York Press, Albany, NY

Fox MW, Bekoff M 1975. The behaviour of dogs. In The Behaviour of Domestic Animals 3rd Edition Hafez ESE (ed), 370-409. Baillière Tindall, London, UK

Fox MW, Stelzner D 1967. The effects of early experience on the development of inter and intraspecies social relationships in the dog. Animal Behaviour 15, 377-386

Freedman DG, King JA, Elliot O 1961. Critical period in the social development of dogs. Science 133, 1016-1017

Fuller JL 1967. Experiential deprivation and later behaviour. Science 158, 1645

Gamble MR 1982. Sound and its significance for laboratory animals. Biological. Reviews 57, 395-421

Gärtner K, Baumans V, Brain PF, Hackbarth H, Militzer K, Morton DB, Nebendahl K, Netto J, Poole TB, Whittaker D 1994. Dogs. In The Accommodation of Laboratory Animals in Accordance with Animal Welfare Requirements. Proceedings of an International Workshop held at the Bundesgesundheitsamt, Berlin [The Berlin Workshop] O'Donoghue PN (ed), 39-46. Bundesministerium für Ernährung, Landwirtschaft und Forsten, Bonn, Germany

Hetts S, Clark JD, Calpin JP, Arnold CE, Mateo JM 1992. Influence of housing conditions on beagle behaviour. Applied Animal Behaviour Science 34, 137-155

Hickey TE 1993. Group housing dogs on GLP toxicology studies. In Refinement and Reduction in Animal Testing Niemi SM, Willson JE (eds), 73-77. Scientists Center for Animal Welfare (SCAW), Bethesda, MD

Home Office 1989. Animals (Scientific Procedures) Act 1986. Code of Practice for the Housing and Care of Animals Used in Scientific Procedures. Her Majesty's Stationery Office, London, UK
Full Text: http://www.homeoffice.gov.uk/animact/hcasp.htm

Hubrecht RC, Serpell JA, Poole TB 1992. Correlates of pen size and housing conditions on the behaviour of kennelled dogs. Applied Animal Behaviour Science 34, 365-383

Hubrecht RC 1993. A comparison of social and environmental enrichment methods for laboratory housed dogs. Applied Animal Behaviour Science 37, 345-361

Hubrecht RC 1995a. Enrichment in puppyhood and its effects on later behavior of dogs. Laboratory Animal Science 45, 70-75

Hubrecht RC 1995b. Housing Husbandry and Welfare Provision for Animals Used in Toxicology Studies: Results of a UK Questionnaire on Current Practice (1994) Universities Federation for Animal Welfare, Potters Bar, UK

Knol BW 1989. Influence of stress on the motivation for agonistic behaviour in the male dog: Role of the hypothalamus pituitary testis system Ph.D. Thesis, University of Utrecht, Utrecht, Netherlands

Loveridge GG 1994. Provision of environmentally enriched housing for dogs. Animal Technology 45, 1-19

Loveridge GG 1998. Environmentally enriched dog housing. Applied Animal Behaviour Science 59, 101-113

MacArthur JA 1999. The Dog. In The UFAW Handbook on the Care and Management of Laboratory Animals 7th edition Poole T (ed), 423-444. Blackwell Science, London, UK

Macdonald DW, Carr GM 1995. Variation in dog society: between resource dispersion and social flux. In The Domestic Dog: Its Evolution, Behaviour, and Interactions with People Serpell J (ed), pp 199-216. Cambridge University, Cambridge, UK

Markwell PJ Thorne CJ 1987. Early behavioural development of dogs. Journal of Small Animal Practice 28, 984-991

Milligan SR, Sales GD, Khirnykh K 1993. Sound levels in rooms housing laboratory animals: an uncontrolled daily variable. Physiology and Behaviour 53, 1067-1076

Ottewill D 1968. Planning and design of accommodation for experimental dogs and cats. Laboratory Animal Symposia 1, 97-112

Petersen EA 1980. Noise and laboratory animals. Laboratory Animal Care 13, 340-350

Sales G, Hubrecht R, Peyvndi A, Milligan S, Shield B 1997. Noise in dog kennelling: Is barking a welfare problem for dogs? Applied Animal Behaviour Science 52, 3-4 321-329

Sales GD, Wilson KJ, Spencer KEV, Milligan SR 1988. Environmental ultrasound in laboratories and animal houses: A possible cause for concern in the welfare and use of laboratory animals. Laboratory Animals 22, 369-375

Scott JP Fuller JL 1965. Genetics and the Social Behaviour of the Dog University of Chicago Press, Chicago, IL

Serpell J (ed) 1995. The Domestic Dog: Evolution, Behaviour, and Interactions with People Cambridge University Press, Cambridge, UK

Thompson WR, Melzack R, Scott TH 1956. "Whirling behaviour" in dogs as related to early exposure. Science 123, 393

Trussell BA, King J, Smith D 1999. Application of environmental enrichment routines to regulatory toxicolgy studies in the Beagle dog. Animal Technology 50, 131-133

United States Department of Agriculture 1991. Title 9, CFR (Code of Federal Register), Part 3. Animal Welfare; Standards; Final Rule. Federal Register 56(32), 6426-6505
Full Text: http://www.nal.usda.gov/awic/legislat/awadog.htm

University of Florida 2000. Guidelines for Exercising Dogs University of Florida—Animal Care & Use, Gainesville, FL
Full Text: http://nersp.nerdc.ufl.edu/~iacuc/dogexec.htm

van der Heiden CV 1992. The problem of noise within kennels: what are its implications and how can it be reduced? The Veterinary Nursing Journal 7, 13-16

van Raaij MTM, Oortgiesen M 1996. Noise stress and airway toxicity: a prospect for experimental analysis. Food and Chemical Toxicology 34, 1159-1161

Wolfle TL 1987. Control of stress using non-drug approaches. Journal of the American Veterinary Medical Association 191, 1219-1221

Wolfle TL 1990. Policy, program and people: The three p's to well-being. In Canine Research Environment Mench JA Krulisch L (eds), 41-47. Scientists Center for Animal Welfare, Bethseda, MD

Wright JC 1980. The development of social structure during the primary socialization period in German Shepherds. Developmental Psychobiology 13, 17-24

Dr. Robert C. Hubrecht is Deputy Director at The Universities Federation for Animal Welfare (UFAW) in Hertfordshire, United Kingdom. UFAW is the key organization in Europe for setting ethical and scientifically sound standards for laboratory housing and husbandry. Robert Hubrecht is an ethologist with comprehensive research and practical experience in the assessment of species-specific housing requirements of dogs.

Comfortable Quarters for Primates in Research Institutions

Viktor Reinhardt

Animal Welfare Institute, PO Box 3650, Washington, DC 20007, USA

In the United States there are approximately 57,000 nonhuman primates kept in laboratories (United States Department of Agriculture, 2000). To safeguard their well-being and their suitability for valid scientific research the following primate-specific characteristics must be taken into consideration in the care and use of these animals:

- Like human primates, nonhuman primates have "social needs" which "must" be addressed in the housing arrangements of animals used in research, testing and experimentation (United States Department of Agriculture, 1991, p. 6499). When they are deprived of companionship for an extended time primates develop unmistakable signs of depression and frustration (Figure 1). Being deprived of social companionship is so distressing that approximately 10 out of 100 single-caged monkeys develop the serious behavioral pathology of self-injurious biting (Jorgensen et al., 1998; Novak et al., 1998; Macy et al, 1999).
- All species, even those who spend much of the day on the ground, seek elevated places as refuge from ground predators and as resting sites for the night (National Research Council, 1998). The presence of large trees or cliffs is often the only limitation for the distribution of nonhuman primates in the wild. When being kept in low-level cages, the animals are cornered and "might perceive the presence of humans above them as particularly threatening" (National Research Council, 1998, p. 118). "Two tier caging should [therefore] be avoided" (Baskerville, 1999).
- Primates are physiologically and anatomically adapted to live in a complex, dynamic environment. Any healthy primate becomes apathetic or restless and develops stereotypical behaviors when kept in surroundings that lack basic species-adequate stimulation. "Psychological well-being is enhanced by…opportunities to engage in behavior related to foraging, exploration, and other activities appropriate to the species" (National Research Council, 1998, p. 2).
- With the exception of one species, primates are diurnal animals. Therefore, "lighting <u>must</u> [emphasis added by author] be uniformly diffused throughout animal facilities and provide sufficient illumination to aid in maintaining

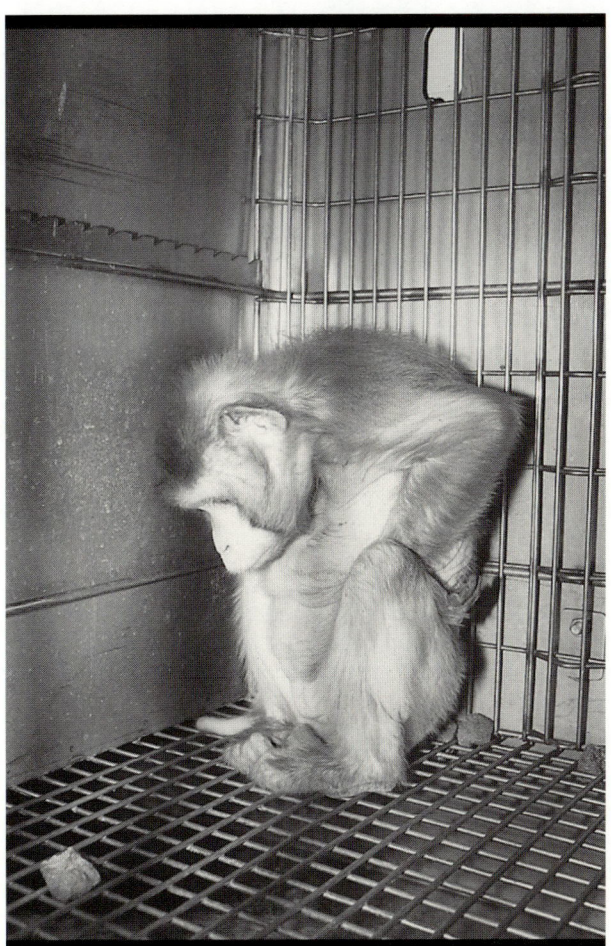

Figure 1. "There is no such thing as a boring animal, rhesus or otherwise; only the conditions under which we keep them can be boring" (Mahoney, 1992, p. 36). Primates have intensive social needs; solitary confinement in a barren cage is a severe punishment for them (Sokol, 1993).

Comfortable Quarters for Laboratory Animals Reinhardt V, Reinhardt A (eds), 65-77. Animal Welfare Institute, Washington, DC 20007

Figure 2. Being housed with a compatible companion—here two male long-tailed macaques engrossed in social grooming—is a good compromise for addressing the social needs of primates kept in research facilities (photo by Richard Lynch, AstraZeneca, Wilmington).

good housekeeping practices, adequate cleaning, adequate inspection of animals, and for the well-being of animals" (United States Department of Agriculture, 1991, p. 6497).

- Primates show physiological and behavioral distress reactions when exposed to life-threatening situations over which they have no control. When this circumstance occurs during a research-related procedure—e.g., enforced immobilization during sample collection—the validity of the findings will be jeopardized because stress has been introduced as a data-biasing variable (Brockway et al., 1993). This problem is usually dealt with by increasing the number of experimental subjects so that data variability decreases and statistical significance can be achieved.

There are simple ways to provide captive primates with housing and handling conditions that address these five characteristics, thereby promoting both the well-being of the animals and the quality of research conducted.

Housing primates in groups is the ideal way to account for their social disposition. Conditions for long-term harmonious group-housing, however, are usually not met in research institutions. Aggression-related injury and social distress can often not be avoided when new groups are formed from previously single-housed individuals. The risk remains inherent even in established groups because of the spatial constraints set by confinement and the artificial instability of social relationships resulting from research and managerial stipulations and from health care considerations. It should be emphasized that these risks can be overcome in research facilities committed to exceptionally high animal husbandry standards (Hartner et al., 2001). **Housing the animals in pairs** offers a practicable alternative to group-housing. Protocols for the safe formation of compatible pairs of previously single-caged adults of the same sex have been documented for numerous species, including chimpanzees (Fritz and Fritz, 1979), baboons (Jerome and Szostak, 1987), squirrel monkeys (Gwinn, 1996), owl monkeys (Weed and Watson, 1998), marmosets (Jackson, 2001; Majolo et al., 2001) rhesus macaques (Reinhardt, 1989a), stump-tailed macaques (Reinhardt, 1994a), long-tailed macaques (Lynch, 1998), and pig-tailed macaques (Byrum and St. Claire, 1998). While adult animals need to be carefully familiarized before being introduced with each other in the same cage, there is little or no danger involved when new pairs are formed by introducing juvenile individuals with strange single-housed adults or by introducing unfamiliar juveniles with each other (Reinhardt, 1994b). Choice studies conducted with capuchin monkeys have demonstrated that the animals "value social companionship as they value food: It is a necessity, not a luxury" (Dettmer and Fragaszy, 2000, p. 303). The social companion is the one environmental enrichment option of which the caged primate never gets bored (cf., de Waal, 1992). Partners of compatible pairs spend about the same amount of time interacting with each other as wild animals do (Line et al., 1990; Reinhardt, 1990a; Brent, 1992a) "even" when they have lived together for several years (Ranheim and Reinhardt, 1989). This suggests that pair-housing is a reasonable option for addressing their social needs

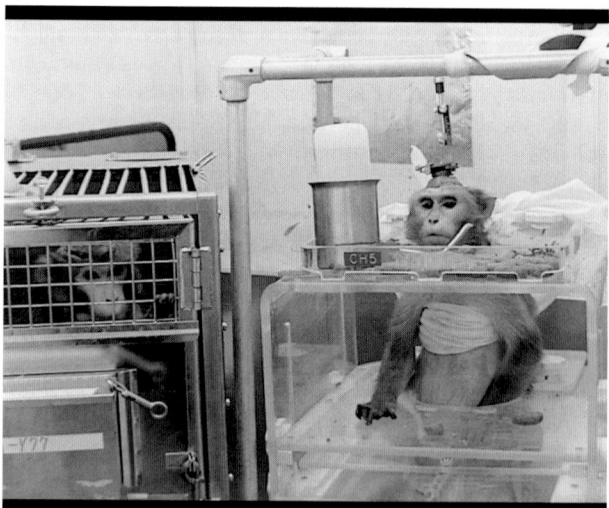

Figure 3. During potentially distressing experimental situations there should always be a compatible companion close by, serving as a psychological support.

(Figure 2). The behavioral disorder of self-injurious biting is resistant to occupational therapy attempts (Preilowski et al., 1988; Kinsey et al., 1997; Novak et al., 1998) but can be cured by transferring the afflicted single-caged individuals to a compatible pair-housing setup (Line et al., 1990; Bushong et al., 1992; Reinhardt, 1999). Transferring single-caged individuals to compatible pair-housing not only improves their behavioral health, but it also decreases their susceptibility to physical health problems (Schapiro and Bushong, 1994), particularly to stress-related diseases (Schapiro et al., 2000a). The two partners of a compatible pair do not differ from each other in terms of stress-sensitive parameters (Gonzalez et al., 1982; Reinhardt et al., 1991; Eaton et al., 1994). Rather than being a source of distress, the companion serves as a stress buffer in fear-provoking situations such as experiments requiring chair-restraint or tether-restraint (Mason, 1960; Coe et al., 1982; Coelho et al., 1991; Figure 3).

Figure 4. Pair-housing does not jeopardize in any way the health status of animals with headcap implants.

The horizontally arranged double cage can provide a suitable primary enclosure for paired primates. It must be furnished with two feeders and two elevated resting sites, one in each half of the cage. A dividing panel with a passage hole close to the back wall of the cage minimizes the risk of squabbles over access to these essential resources and makes it possible for the two animals to sometimes get out of each other's sight. This avoids antagonism while fostering affiliation (Reinhardt and Reinhardt, 1991). The vertically arranged double cage does not offer suitable housing conditions because the dominant partner may monopolize the well-illuminated environment of the upper half thereby forcing the subordinate partner to spend most of the time in the dim environment of the bottom half (own unpublished observation; cf., Williams et al., 1988). Housing primates in pairs does not interfere with husbandry procedures and common management and research protocols (Reinhardt et al., 1989; Schapiro and Bushong, 1994; Shively, 2001; Figure 4). If an animal has to be kept temporarily alone—e.g., during intensive postoperative care or during a metabolic study—the caging arrangement must allow the individual animal to maintain visual contact with at least one compatible conspecific (cf., Mahoney, 1992; Lindburg and Coe, 1995) to minimize the stress resulting from social deprivation (cf., Coelho et al., 1991). "We all realize that one is better off with the ups and downs of a social life than without a social life at all. Would it be any different with nonhuman primates?" (de Waal, 1992, p. 86).

While social contact and social interaction with another compatible conspecific is a prerequisite for the psychological and behavioral health of a primate, a **trustful relationship with the attending personnel** is essential for his/her well-being. Giving the animals names, rather than referring to them as numbered research objects helps to develop such a relationship (cf., Reese, 1991; Sokol, 1992; de Waal, 1992). The compassionate and respectful attitude "conveys to the animal a quiet sense of assurance on which coping strategies can be developed for dealing with other stressful aspects of the laboratory" (Wolfle, 1987, p. 1221). The macho-type person is out of place in the animal room because s/he triggers distress reactions which will skew scientific findings even before the actual experiment has started. Typically, the animals will panic when such a person comes into their room (own unpublished observation; cf., Arluke and Sanders, 1996). In order to provide not only a physically but also a psychologically comfortable environment, staff time must be set aside for interacting and communicating with the animals in a positive manner every day (cf., Home Office, 1989). "There should be no sharp demarcation between 'good guys' and 'bad guys.' Nonhuman primates are quick to forget, or perhaps forgive, the momentary fear or resentment they feel towards a human being who has just subjected them to an unpleasant experience if a strong bond of trust already exists with that person" (Mahoney, 1992, p. 35). The investigator, veterinarian or technician who pretends to be too busy to deal with the animals other than during experimental or clinical procedures lacks basic professional qualification.

Indoor enclosures of primates should take maximum advantage of the height of the room, allowing the installation of climbing structures and resting surfaces at different levels. **Elevated structures** not only increase the usable cage volume but also provide environmental enrichment by stimulating the animals to spend a major portion of their time engaged in species-typical arboreal activities. A perch can readily be installed in any cages, "even" in those that are equipped with a movable back-wall (Reinhardt and Pape, 1991; Watson, 1991). The monkey perch should be made of non-metallic, i.e., "warm" material, and have a sufficiently large diameter so that the animal(s) can sit on it comfortably (Abee, 1985) or, as in the case of squirrel monkeys and lemurs, lie flat on it with limbs dangling on either side for balance (own unpublished observation; McGivern, 1993). The perch must be placed at a height that the animal(s) are able to sit on and under it in an unrestricted, i.e, not crouched position (Reinhardt and Reinhardt, 1999). Monkeys prefer to look out of their enclosure rather than hide in the back of it (van Wagenen, 1950; Reinhardt, 1989b). Elevated fixtures, therefore, have to be placed in such a way that the animal(s) can sit on them in a look-out position in the front of the cage (Figure 5). An adequate number of elevated sites must be provided for

Figure 5. The elevated resting surface should be placed in such a way that the animal can sit on it in a look-out position.

Figure 6. Placing the food on the mesh ceiling of the cage is an inexpensive but very effective way of allowing primates to engage in foraging activities.

Kaumanns and Schönmann, 1997; Wakenshaw, 1999). Mobile structures such as swings and ropes are generally less preferred than fixed structures (Williams et al., 1988; Howell et al., 1997), particularly in cages that provide only the bare minimum space for postural adjustment (Kopecky and Reinhardt, 1991).

Since primates are biologically programmed to spend the major portion of their time gathering and processing food, feeding enrichment provides optimal **environmental stimulation** for them. The urge to forage is so strong that they will work for food even when identical food is placed right next to them (Markowitz, 1979; Evans et al., 1989; Reinhardt, 1994c). Distributing the food on the chain-link or mesh ceiling of the enclosure rather than in food boxes, troughs or on the floor is perhaps the simplest way to trigger foraging behavior for standard food and supplemental food, such as fruits, vegetables and bread (Figure 6). Depending on the size of the individual food pieces, macaques increase their foraging time more than 100-fold when their daily standard biscuit ration is placed on the mesh ceiling of their cage rather than in ordinary, freely accessible food boxes (Reinhardt, 1993a). The same effect can be achieved when the food box is remounted a few centimeters away from its access hole (Reinhardt, 1993b). Dexterity is now required to maneuver the biscuits through the mesh covering the face of the box (Figure 7). These two feeding enrichment options use structural elements of the cage, redesigned in such a way that they serve as primary feeders for the daily standard food ration. Therefore, no extra time is needed to clean them and bait them with special treats. The animals only work for food that they actually eat (Murchison, 1994; Reinhardt and Garza-Schmidt, 2000). This avoids the accumulation of spoiled foodstuff, a problem which commonly occurs with ordinary feeders, when the animals hoard food and waste part of it by dropping it on fecal material. Marmosets and tamarins are adapted to probe for embedded food in order to extract it. Simulating a natural food-source with an artificial device stocked with gum (McGrew and Brennan, 1986; König et al., 1987; Kelly, 1993) or baited with raisins mixed with corn cob (Molzen and French, 1989) is cheap and easy but allows the animals to engage in extensive foraging activities (Figure 8). Chimpanzees are proficient in using twigs to extract termites from logs and mounds. Simple probefeeder devices filled with sticky foods—such as applesauce, mashed bananas and jam—can encourage this mode of foraging in captivity (Murphy, 1976; Goodall, 1979; Nash, 1982; Maki et al., 1989; Gilloux et al., 1992; Perret et al., 1998; Figure 9). Straw, woodchips, woodwool, and argilla espansa are ideal foraging substrates in the form of deep-litter on the solid floor of pens (Chamove and Anderson, 1979; Fragaszy and Adams-Curtis, 1991; Brent, 1992b; Beck, 1995; Brown et al., 1995; Riviello and Misiti, 1995; Baker, 1997) or distributed on trays mounted under the mesh floor of cages (Bryant et al., 1988; Mahoney, 1992). When mixed with seeds, grain, or other small edible items such materials promote intensive foraging behavior (Burt and Plant, 1990; Byrne and Suomi, 1991; Combette and Anderson, 1991; Grief et al., 1992; Poenisch, 1992; Figure 10). The distracting effect of this feeding enrichment technique is so strong that it reduces social antagonism in group-housed animals (Chamove et al., 1982; Boccia, 1989).

socially housed animals to avoid competition over access (Williams et al., 1988). When this condition is met, access to perches helps the animals to avoid social conflicts and foster affiliative relationships (Neveu and Deputte, 1996). Nonhuman primates can get extremely frightened when a person dressed in protective garb approaches them. To minimize the stress, there should be sufficiently high structures that allow the animals to show a vertical flight response and retreat to a quasi safe location above human eye level (cf., International Primatological Society, 1993;

Figure 7. The primary feeder can be transferred into a foraging device by remounting it away from the access hole. Dexterity is now required to maneuver the biscuits through the mesh covering the face of the box.

Figure 9. In the wild, chimpanzees are proficient in extracting termites from logs and mounds. Probefeeders can promote this behavior in the research laboratory (photo by Kai Perret, Allwetterzoo Münster, Germany).

Figure 8. In their natural environment marmosets show "tree-gouging" and "exudate-feeding." A simple gum-feeder can support this behavior under laboratory conditions (photo by Katie Eckert, University of California, San Francisco, CA).

Figure 10. Woodchips provide an ideal foraging substrate for group-housed primates; here a group of stump-tailed macaques foraging for seeds (photo by James R. Anderson, University of Sterling, UK).

All captive primates should be fed daily at least one whole, medium-size fruit or vegetable (Figure 11a,b,c). The time required by animal care personnel to cut produce into small pieces can be spent in much more meaningful ways (cf., Smith et al., 1989; Kerridge, 1997). For example, portioning the daily standard food ration in several rather than a single feeding probably does not take more time than chopping the supplemental produce for the animals, but it increases substantially the amount of time they can engage in foraging activities (cf., Taylor et al., 1997).

Figure 11a,b,c. Primates deserve fresh fruit or vegetables on a daily basis. It would be a waste of time to chop the produce for the animals; they have the time and they enjoy doing it themselves.

Branch segments of dead deciduous trees—red oak disintegrates into flakes that are so small that large quantities pass sewage drains without clogging them (Reinhardt, 1992a)—are perfect toys for primates, stimulating not only processing but also manipulative and play activities (Figure 12). They constantly change their form and texture due to wear and dehydration and, therefore, retain their stimulatory value (Reinhardt, 1989b; Eckert et al., 2000). Commercial toys lack the natural, ever-changing texture of wood; this is probably the reason why the animals quickly lose interest in them (Crockett et al., 1989; Line et al., 1989; Hamilton, 1991; Pruetz and Bloomsmith, 1992; Kessel and Brent, 1998), unless several different toys are offered and substituted regularly with new ones (cf., Paquette and Prescott, 1988; Weick et al., 1991). Access to a variety of manipulable objects seems to be particularly beneficial for capuchins and baboons, who exhibit sustained interest in them and respond with a significant reduction in abnormal behaviors (Brent and Belik, 1997; Boinski et al., 1999).

A major contention is the need for proper **illumination** in the caging arrangement of medium- and small-sized primates. In order to minimize housing expenses, these animals are traditionally kept in two-tier cages, with one row stacked on top of another. This doubles the number of primates that can be accommodated in one room, but involves serious implications for the individual animals. Those relegated to the lower rows are restricted to a terrestrial lifestyle, unable to withdraw in alarming situations and retreat to a safe place above the human predators who periodically capture them and subject them to distressing, or even deadly procedures. Moreover, the sanitation tray, which runs the length of the room beneath the upper tier of cages, reduces significantly the amount of light that can penetrate to the lower-cage tier (Schapiro et al., 2000b); "animals in the lower tier are thus relegated to a permanent state of semi-gloom" (Mahoney, 1992; p. 32). The cave-like living quarters of bottom-row caged animals is often so dim that caretakers routinely have to use flashlights to identify and inspect them (Figure 13). It has been noticed in marmosets that the housing environment of lower-row caged animals can be so poor that it results in markedly reduced fertility (Heger et al., 1986). Routinely rotating animals between bottom and top tiers (National Research Council, 1998) offers no solution to this problem. It merely "rotates" the problem by alleviating the situation for lower-row subjects, while aggravating it for the same number of upper-row subjects. At the same time it introduces the additional stress-variable associated with cage transfer (Mitchell and Gomber, 1976; Phoenix and Chambers, 1984; Crockett et al., 1993; Schapiro et al., 1997). Even if techniques can be developed to assure uniform illumination, the bottom-tier cage will remain a potential source of distress whenever personnel enters the room (cf., Kaumanns and Schönmann, 1997). In order to provide ethically and scientifically acceptable caging conditions, nonhuman primates must be housed in single-row cages to assure that (a) all animals receive the same quantity and quality of light, (b) all cages

Figure 12. Branch segments are perfect toys for primates. Constantly changing their texture and configuration due to wear, these wooden toys do not lose their stimulatory value over time.

are of sufficient height so that occupants are in a position to retreat above animal care personnel, and (c) all animals in the room can be adequately inspected.

Training nonhuman primates to cooperate during procedures is one of the most significant options of making life a little bit more bearable for laboratory primates. It not only challenges the animals' high degree of intelligence, offers them—and the caregivers—some relevant distraction and eliminates data-confounding distress responses, but it also increases personnel safety by no longer giving the animals reason to defend themselves by means of biting or scratching during compulsory immobilization. To be forcefully removed from the familiar cage and subdued during painful husbandry and research procedures must, indeed, be a terribly frightening experience for a monkey or an ape. Research data collected from such an animal are tainted by the subject's stress reactions (review: Reinhardt et al., 1995) and, therefore, have questionable scientific value (Figure 14). With gentle firmness, patience and positive reinforcement many primate species can be conditioned to work with—rather than against—personnel during common procedures such as transfer to a holding area (Goodwin, 1997; Bloomsmith et al., 1998), capture from the home cage (Reinhardt, 1992b), capture from the group (Reinhardt, 1990b,c; Kessel-Davenport and Gutierrez, 1994; Mendoza, 1999; White et al., 2000), blood collection (McGinnis and Kraemer, 1979; Reinhardt, 1991; Laule et al., 1996; Moore and Suedmeyer, 1997), blood pressure measurement (Smith and Ansevin, 1957; Mitchell et al., 1980; Turrkan, 1990), systemic injection (Spragg, 1940; Levison et al., 1964; Byrd, 1977; Priest, 1991; Reinhardt, 1992c; Figure 15a,b), urine collection (Kelly and Bramblett, 1981; Ziegler et al., 1987; Bond, 1991; Anzenberger and Gossweiler, 1993; Shideler et al., 1994), saliva collection (Bettinger, 1998; Bettinger et al., 1998), topical drug application (Reinhardt and Cowley, 1990; Segerson and Laule, 1995), oral drug application (Turrkan et al., 1989), semen collection (Brown and Loskutoff, 1998), insemination (Desmond et al., 1987), vaginal swabbing (Bunyak et al., 1992; Hernándes-López et al., 1998) and veterinary examination (Brown, 1998). The initial time investment in the training quickly pays off in: (a) a reduction of time required to obtain a sample, administer a drug or capture an animal, (b) a reduction of risks associated with defense aggression, (c) a reduction in the use of pharmacological restraint agents, (d) more reliable research data (Elvidge et al., 1976; Reinhardt, 1992c; Schnell and Gerber, 1997; National Research Council, 1998) and a more satisfactory relationship between handling personnel and research subject (Figure 16a,b,c,d).

Concluding Remarks

Providing primates in research institutions with primate-adequate housing and humane handling conditions is no sentimentalism. On the contrary, it is essential to employ such refined methodology in order to adhere to the very basic principles of good science. A primate who behaves like a primate and who is free of

Figure 13. Even though it is the prevailing housing arrangement, the double-tier caging system is unacceptable both for ethical and scientific reasons.

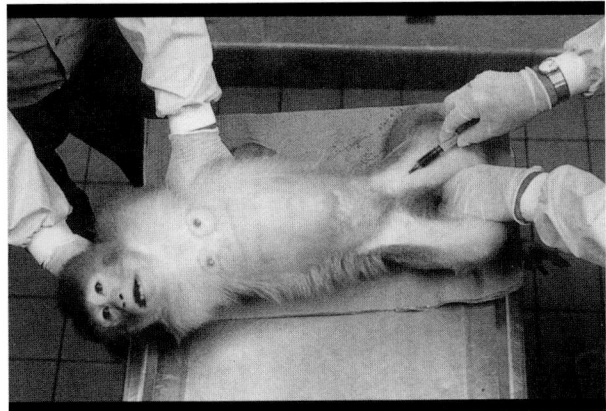

Figure 14. Scientific data collected from a subdued animal are skewed by the subject's fear response.

 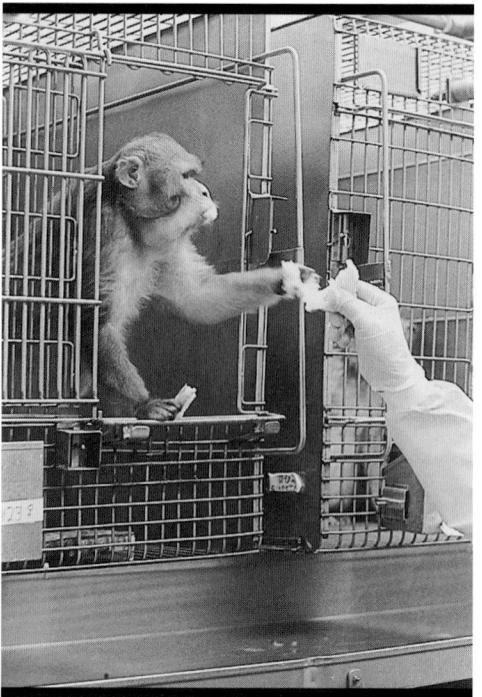

Figure 15a,b. There is no need to forcefully restrain an animal for Ketamin injection, thereby introducing stress as an uncontrolled variable even before the actual experiment has started. Primates are intelligent and can be readily trained to cooperate during such a simple procedure.

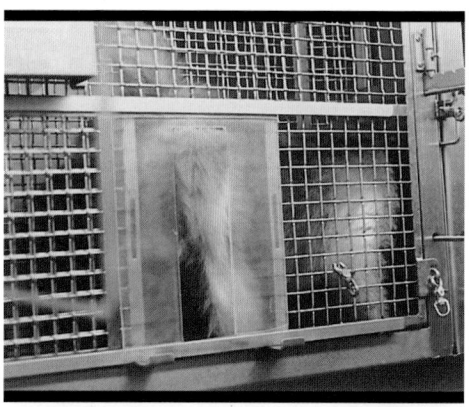 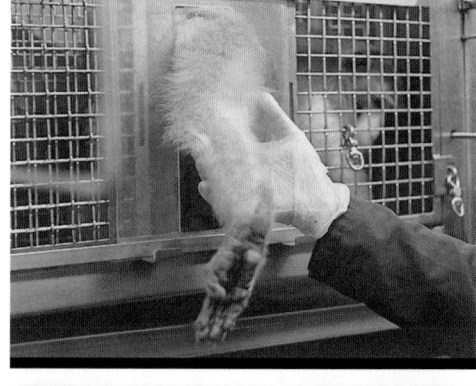
 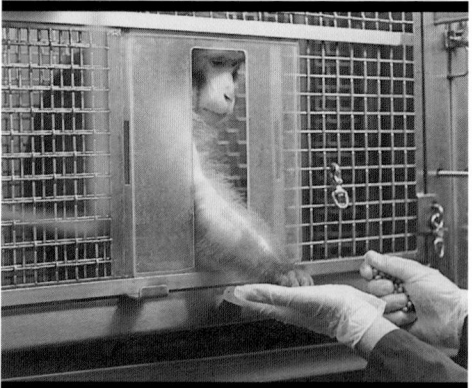

Figure 16a,b,c,d. It takes less than a cumulative total of one hour to train adult male rhesus macaques to cooperate—rather than resist—during blood collection in the familiar home cage (Reinhardt, 1991).

distress, certainly has a greater potential of being a useful research model than one who is a behavioral cripple as a result of understimulation and one who experiences distress during research-related procedures.

References

Abee CR 1985. Medical care and management of the squirrel monkey. In Handbook of Squirrel Monkey Research Rosenblum LA, Coe CL (eds), 447-488. Plenum Press, New York, NY

Anderson JR, Chamove AS 1984. Allowing captive primates to forage. In Standards in Laboratory Animal Management. Proceedings of a Symposium, 253-256. The Universities Federation For Animal Welfare, Potters Bar, UK
Full Text: http://www.awionline.org/Lab_animals/biblio/ufaw-2~1.htm

Anzenberger G, Gossweiler H 1993. How to obtain individual urine samples from undisturbed marmoset families. American Journal of Primatology 31, 223-230

Arluke A, Sanders CR 1996. Regarding Animals. Temple University Press, Philadelphia, PA

Baker KC 1997. Straw and forage material ameliorate abnormal behaviors in adult chimpanzees. Zoo Biology 16, 225-236

Baskerville M 1999. Old World Monkeys. In The UFAW Handbook on the Care and Management of Laboratory Animals Seventh Edition Poole T, English P (eds), 611-635. Blackwell Science, Oxford, UK

Beck RPA 1995. A study of environmental enrichment in groups of captive lion tamarins (*Leontopithecus rosalia* & *Leontopithecus chrysmelas*). RATEL (Journal of the Association of British Wild Animal Keepers) 22(4), 112-126

Bettinger T 1998. Saliva Collection of Trained Adult Male Gorillas (Videotape with commentary). In Workshop—Advances in Primate Training. Twenty-first Annual Meeting of the American Society of Primatologists. O'Neill-Wagner P, Stone A (eds). Cleveland Metroparks Zoo, Cleveland, OH

Bettinger T, Kuhar C, Sironen A, Laudenslager M 1998. Behavior and salivary cortisol in gorillas housed in an all male group. American Zoo and Aquarium Association (AZA) Annual Conference Proceedings, 242-246

Bloomsmith MA, Stone AM, Laule GE 1998. Positive reinforcement training to enhance the voluntary movement of group-housed chimpanzees within their enclosure. Zoo Biology 17, 333-341

Boccia ML 1989. Long-term effects of a natural foraging task on aggression and stereotypies in socially housed pigtail macaques. Laboratory Primate Newsletter 28(2), 18-19
Full Text: http://www.brown.edu/Research/Primate/lpn28-2.html#maria

Bond M 1991. How to collect urine from a gorilla. Gorilla Gazette 5(3), 12-13

Boinski S, Swing SP, Gross TS, Davis JK 1999. Environmental enrichment of brown capuchins (*Cebus apella*): Behavioral and plasma and fecal cortisol measures of effectiveness. American Journal of Primatology 48, 49-68

Brent L 1992a. The effects of cage size and pair housing on the behavior of captive chimpanzees. American Journal of Primatology 27, 20

Brent L 1992b. Woodchip bedding as enrichment for captive chimpanzees in an outdoor enclosure. Animal Welfare 1, 161-170
FT: http://www.awionline.org/Lab_animals/biblio/aw1-1.htm

Brent L, Belik M 1997. The response of group-housed baboons to three enrichment toys. Laboratory Animals 31, 81-85
Full Text: http://www.lal.org.uk/pdf.htm

Brockway BP, Hassler CR, Hicks N 1993. Minimizing stress during physiological monitoring. In Refinement and Reduction in Animal Testing Niemi SM, Willson JE (eds), 56-69. Scientists Center for Animal Welfare, Greenbelt, MD

Brown CS 1998. A Training Program for Semen Collection in Gorillas (Videotape with commentary). Omaha's Henry Doorly Zoo, Omaha, NE

Brown CS, Loskutoff NM 1998. A training program for noninvasive semen collection in captive western lowland gorillas (*Gorilla gorilla gorilla*). Zoo Biology 17, 143-151

Brown DL, Calcagno J, Gold KC, Thompson S 1995. Effects of environmental enrichment on nonsocial and abnormal behavior of captive lowland gorillas (*Gorilla gorilla gorilla*). American Zoo and Aquarium Association (AZA) Regional Conference Proceedings, 29-35

Bryant CE, Rupniak NMJ, Iversen SD 1988. Effects of different environmental enrichment devices on cage stereotypies and autoaggression in captive cynomolgus monkeys. Journal of Medical Primatology 17, 257-269
FT: http://www.awionline.org/Lab_animals/biblio/jmp17-2.htm

Bunyak SC, Harvey NC, Rhine RJ, Wilson MI 1982. Venipuncture and vaginal swabbing in an enclosure occupied by a mixed-sex group of stumptailed macaques (*Macaca arctoides*). American Journal of Primatology 2, 201-204

Burt DA, Plant M 1990. Observations on a caging system for housing stump-tailed macaques. Animal Technology 41, 175-179
FT: http://www.awionline.org/Lab_animals/biblio/at-burt.htm

Bushong D, Schapiro SJ, Bloomsmith MA 1992. Self-aggression in nonhuman primates: A review of its development/possible causes, methods of therapeutic treatment, and its relevance to the zoo situation. American Zoo and Aquarium Association (AZA) Regional Conference Proceedings, 723-728

Byrd LD 1977. Introduction: Chimpanzees as biomedical models. In Progress in Ape Research Bourne GH (ed), 161-165. Academic Press, New York. NY

Byrne GD, Suomi SJ 1991. Effects of woodchips and buried food on behavior patterns and psychological well-being of captive rhesus monkeys. American Journal of Primatology 23, 141-151

Byrum R, St. Claire M 1998. Pairing female *Macaca nemestrina*. Laboratory Primate Newsletter 37(4), 1
Full Text: http://www.brown.edu/Research/Primate/lpn37-4.html#byrum

Chamove AS, Anderson JR 1979. Woodchip litter in macaque group. Animal Technology 30, 69-74
FT: http://www.awionline.org/Lab_animals/biblio/at-cham.htm

Chamove AS, Anderson JR, Morgan-Jones SC, Jones SP 1982. Deep woodchip litter: Hygiene, feeding, and behavioral enhancement in eight primate species. International Journal for the Study of Animal Problems 3, 308-318
FT: http://www.awionline.org/Lab_animals/biblio/ijsap1.html

Coe CL, Franklin D, Smith ER, Levine S 1982. Hormonal responses accompanying fear and agitation in the squirrel monkey. Physiology and Behavior 29, 1051-1057

Coelho AM, Carey KD, Shade RE 1991. Assessing the effects of social environment on blood pressure and heart rates of baboon. American Journal of Primatology 23, 257-267

Combette C, Anderson JR 1991. Réponses à deux techniques d'enrichissement environmental chez deux espèces de primates en laboratoire (Cebus apella, Lemur macaco). Cahiers d'Ethologie 11, 1-16

Crockett CM, Bielitzki JT, Carey A, Velez A 1989. Kong toys as enrichment devices for singly-caged macaques. Laboratory Primate Newsletter 28(2), 21-22
Full Text: http://www.brown.edu/Research/Primate/lpn28-2.html#kong

Crockett CM, Bowers CL, Sackett GP, Bowden DM 1993. Urinary cortisol responses of longtailed macaques to five cage sizes, tethering, sedation, and room change. American Journal of Primatology 30, 55-74

Desmond T, Laule GM, McNary J 1987. Training to enhance socialization and reproduction in drills. American Zoo and Aquarium Association (AZA) Regional Conference Proceedings, 352-358

Dettmer E, Fragaszy D 2000. Determining the value of social companionship to captive tufted capuchin monkeys (Cebus apella). Journal of Applied Animal Welfare Science 3, 393-304

Eaton GG, Kelley ST, Axthelm MK, Iliff-Sizemore SA, Shiigi SM 1994. Psychological well-being in paired adult female rhesus (Macaca mulatta). American Journal of Primatology 33, 89-99

Eckert K, Niemeyer C, Anonymous, Rogers RW, Seier J, Ingersoll B, Barklay L, Brinkman C, Oliver S, Buckmaster C, Knowles L, Pyle S 2000. Wooden objects for enrichment: A discussion. Laboratory Primate Newsletter 39(3), 1-4
Full Text: http://www.brown.edu/Research/Primate/lpn39-3.html#wood

Elvidge H, Challis JRG, Robinson JS, Roper C, Thorburn GD 1976. Influence of handling and sedation on plasma cortisol in rhesus monkeys (Macaca mulatta). Journal of Endocrinology 70, 325-326

Evans HL, Taylor JD, Ernst J, Graefe JF 1989. Methods to evaluate the well-being of laboratory primates. Comparison of macaques and tamarins. Laboratory Animal Science 39, 318-323

Fragaszy DM, Adams-Curtis LE 1991. Environmental challenges in groups of capuchins. In Primate Responses to Environmental Change Box HO (ed), 247-264. Chapman and Hall, New York, NY

Fritz P, Fritz J 1979. Resocialization of chimpanzees. Journal of Medical Primatology 8, 202-221

Gilloux I, Gurnell J, Shepherdson D 1992. An enrichment device for great apes. Animal Welfare 1, 279-289
FT: http://www.awionline.org/Lab_animals/biblio/aw1-279.htm

Gonzalez CA, Coe CL, Levine S 1982. Cortisol responses under different housing conditions in female squirrel monkeys. Psychoneuroendocrinology 7, 209-216

Goodall J 1979. Anti-boredom devices for primates. In Comfortable Quarters for Laboratory Animals, Seventh Edition, 16. Animal Welfare Institute, Washington, DC

Goodwin J 1997. The application, use, and effects of training and enrichment variables with Japanese snow macaques (Macaca fuscata) at the Central Park Wildlife Center. American Zoo and Aquarium Association (AZA) Regional Conference Proceedings, 510-515

Grief L, Fritz J, Maki S 1992. Alternative forage types for captive chimpanzees. Laboratory Primate Newsletter 31(2), 11-13
Full Text: http://www.brown.edu/Research/Primate/lpn31-2.html#grief

Gwinn LA 1996. A method for using a pole housing apparatus to establish compatible pairs among squirrel monkeys. Contemporary Topics in Laboratory Animal Science 35(4), 61

Hamilton P 1991. Enrichment toys and tools in recent trials. Humane Innovations and Alternatives in Animal Experimentation 5, 272-277

Hartner M, Hall J, Penderghest J, Clark LP 2001. Group-housing subadult male cynomolgus macaques in a pharmaceutical environment. Lab Animal 30(8), 53-57

Heger W, Merker H-J, Neubert D 1986. Low light intensity decreases the fertility of Callithrix jacchus. Primate Report 14, 260
FT: http://www.awionline.org/Lab_animals/biblio/pr14-2.htm

Hernándes-López L, Mayagoitia L, Esquivel-Lacroix C, Rojas-Maya S, Mondragón-Ceballos R 1998. The menstrual cycle of the spider monkey (Ateles geoffroyi). American Journal of Primatology 44, 183-195

Home Office 1989. Animals (Scientific Procedures) Act 1986. Code of Practice for the Housing and Care of Animals Used in Scientific Procedures. Her Majesty's Stationery Office, London, UK
Full Text: http://www.homeoffice.gov.uk/animact/hcasp.htm

Howell SM, Mittra E, Fritz J, Baron J 1997. The provision of cage furnishings as environmental enrichment at the Primate Foundation of Arizona. The Newsletter 9(2), 1-5
FT: http://www.awionline.org/Lab_animals/biblio/jo-5.htm

International Primatological Society 1993. IPS International guidelines for the acquisition, care and breeding of nonhuman primates, Codes of Practice 1-3. Primate Report 35, 3-29
Full Text: http://www.primate.wisc.edu/pin/ips2.html

Jackson MJ 2001. Environmental enrichment and husbandry of the MPTP-treated common marmoset. Animal Technology 52, 21-28

Jerome CP, Szostak L 1987. Environmental enrichment for adult, female baboons (Papio anubis). Laboratory Animal Science 37, 508-509

Jorgensen MJ, Kinsey JH, Novak MA 1998. Risk factors for self-injurious behavior in captive rhesus monkeys (Macaca mulatta). American Journal of Primatology 45, 187

König A, Rothe H, Radespiel U, Darms K, Bodemeyer J 1987. Nagehölzer für Krallenaffen. Zeitschrift des Kölner Zoo 30(3), 107-108

Kaumanns W, Schönmann U 1997. Requirements for cebids. Primate Report 49, 71-91

Kelley TM, Bramblett CA 1981. Urine collection from vervet monkeys by instrumental conditioning. American Journal of Primatology 1, 95-97

Kelly K 1993. Environmental enrichment for captive wildlife through the simulation of gum feeding. Animal Welfare Information Center Newsletter 4(3), 1-2 & 5-10
FT: http://www.nal.usda.gov/awic/newsletters/v4n3/4n3.htm

Kerridge FJ 1997. Behavioural enrichment of ruffed lemurs (Varecia variegata) based upon a wild-captive comparison of their behaviour. Primate Eye 63, 36-37

Kessel AL, Brent L 1998. Cage toys reduce abnormal behavior in individually housed pigtail macaques. Journal of Applied Animal Welfare Science 1, 227-234

Kessel-Davenport AL, Gutierrez T 1994. Training captive chimpanzees for movement in a transport box. The Newsletter 6(2), 1-2
FT: http://www.awionline.org/Lab_animals/biblio/jo-6.htm

Kinsey JH, Jorgensen MJ, Novak MA 1997. The effects of grooming boards on abnormal behavior in rhesus monkeys (Macaca mulatta). American Journal of Primatology 42, 122-123

Kopecky J, Reinhardt V 1991. Comparing the effectiveness of PVC swings versus PVC perches as environmental enrichment objects for caged female rhesus macaques. Laboratory Primate Newsletter 30(2), 5-6
FT: http://www.brown.edu/Research/Primate/lpn30-2.html#vik

Laule GE, Thurston RH, Alford PL, Bloomsmith MA 1996. Training to reliably obtain blood and urine samples from a diabetic chimpanzee (Pan troglodytes). Zoo Biology 15, 587-591

Levison PK, Fester CB, Nieman WH, Findley JD 1964. A method for training unrestrained primates to receive drug injection. Journal of the Experimental Analysis of Behavior 7, 253-254

Lindburg DG, Coe J 1995. Ark design update: Primate needs and requirements. In Conservation of Endangered Species in Captivity Gibbons EF, Durrant BS, Demarest AJ (eds), 553-570. SUNY Press, Albany, NY

Line SW, Markowitz H, Morgan KN, Strong S 1989. Evaluation of attempts to enrich the environment of single-caged non-human primates. In Animal Care and Use in Behavioral Research: Regulation, Issues, and Applications Driscoll JW (ed), 103-117. Animal Welfare Information Center National Agricultural Library, Beltsville, MD

Line SW, Morgan KN, Markowitz H, Roberts J, Riddell M 1990. Behavioral responses of female long-tailed macaques (Macaca fascicularis) to pair formation. Laboratory Primate Newsletter 29(4), 1-5
Full Text: http://www.brown.edu/Research/Primate/lpn29-4.html#line

Lynch R 1998. Successful pair-housing of male macaques (Macaca fascicularis). Laboratory Primate Newsletter 37(1), 4-5
Full Text: http://www.brown.edu/Research/Primate/lpn37-1.html#pair

Macy JD, Beattie TA, Morgenstern SE, Arnstern AFT 1999. The use of guanfacine to control self-injurious behavior in nonhuman primates. Abstracts of the AALAS [American Association for Laboratory Animal Science] Meeting, 9

Mahoney CJ 1992. Some thoughts on psychological enrichment. Lab Animal 21(5), 27,29,32-37

Majolo B, Buchanan-Smith HM, Morris K 2001. Factors affecting the successful pairing of unfamiliar common marmoset (Callithrix jacchus) females. Primate Eye 73, 12-13
Full Text: http://www.psgb.org/Meetings/Winter2000.html

Maki S, Alford PL, Bloomsmith MA, Franklin J 1989. Food puzzle device simulating termite fishing for captive chimpanzees (Pan troglodytes). American Journal of Primatology 19 (Supplement 1), 71-78

Markowitz H 1979. Environmental enrichment and behavioral engineering for captive primates. In Captivity and Behavior Erwin J, Maple T, Mitchell G (eds), 217-238. Van Nostrand Reinhold, New York, NY

Mason WA 1960. Socially mediated reduction in emotional responses of young rhesus monkeys. Journal of Abnormal and Social Psychology 60, 100-110

McGinnis PR, Kraemer HC 1979. The Stanford outdoor primate facility. In Comfortable Quarters for Laboratory Animals, Seventh Edition, 20-27. Animal Welfare Institute, Washington, DC

McGivern L 1993. Small primate enrichment at the Calgary Zoo, part 2: Lemurs. The Shape of Enrichment 2(4), 9-10

McGrew WC, Brennan JARJ 1986. An artificial 'Gum-tree' for marmosets (Callithrix j. jacchus). Zoo Biology 5, 45-50

Mendoza SP 1999. Squirrel Monkeys. In The UFAW Handbook on the Care and Management of Laboratory Animals Seventh Edition UFAW [Universities Federation for Animal Welfare] Poole T, English P (eds), 591-600. Blackwell Science, Oxford, UK

Mitchell DS, Wigodsky HS, Peel HH, McCaffrey TA 1980. Operant conditioning permits voluntary, noninvasive measurement of blood pressure in conscious, unrestrained baboons (Papio cynocephalus). Behavior Research Methods and Instrumentation 12, 492-498

Mitchell G, Gomber J 1976. Moving laboratory rhesus monkeys (Macaca mulatta) to unfamiliar home cages. Primates 17, 543-547

Molzen EM, French JA 1989. The problem of foraging in captive callitrichid primates: Behavioral time budgets and foraging skills. In Housing, Care and Psychological Wellbeing of Captive and Laboratory Primates Segal EF (ed), 89-101. Noyes Publications, Park Ridge, NJ

Moore BA, Suedmeyer K 1997. Blood sampling in 0.2 Bornean orangutans at the Kansas City Zoological Gardens. Animal Keepers' Forum 24, 537-540

Murchison MA 1994. Primary forage feeder for singly-caged macaques. Laboratory Primate Newsletter 33(1), 7-8
Full Text: http://www.brown.edu/Research/Primate/lpn33-1.html#mark

Murphy DE 1976. Enrichment and occupational devices for orangutans and chimpanzees. International Zoo News 137(23.5), 24-26
Full Text: http://www.awionline.org/Lab_animals/biblio/izn-mur.htm

Nash VJ 1982. Tool use by captive chimpanzees at an artificial termite mound. Zoo Biology 1, 211-221

National Research Council 1998. The Psychological Well-Being of Nonhuman Primates. National Academy Press, Washington, DC
Full Text: http://pompeii.nap.edu/books/0309052335/html/index.html

Neveu H, Deputte BL 1996. Influence of availability of perches on the behavioral well-being of captive, group-living mangabeys. American Journal of Primatology 38, 175-185

Novak MA, Kinsey JH, Jorgensen MJ, Hazen TJ 1998. Effects of puzzle feeders on pathological behavior in individually housed rhesus monkeys. American Journal of Primatology 46, 213-227

Paquette D, Prescott J 1988. Use of novel objects to enhance environments of captive chimpanzees. Zoo Biology 7, 15-23

Perret K, Büchner S, Adler HJ 1998. Beschäftigungsprogramme für Schimpansen (Pan troglodytes) im Zoo. (Environmental enrichment program for chimpanzees in zoos) [German text with English summary]. Der Zoologische Garten 68, 95-111

Phoenix CH, Chambers KC 1984. Sexual behavior and serum hormone levels in aging rhesus males: Effects of environmental change. Hormones and Behavior 18, 206-215

Poenisch T 1992. Bedding for enrichment. The Newsletter 4(1), 1
  FT: http://www.awionline.org/Lab_animals/biblio/jo-14.htm

Preilowski B, Reger M, Engele H 1988. Combining scientific experimentation with conventional housing: A pilot study with rhesus monkeys. American Journal of Primatology 14, 223-234

Priest GM 1991. Training a diabetic drill (*Mandrillus leucophaeus*) to accept insulin injections and venipuncture. Laboratory Primate Newsletter 30(1), 1-4
  Full Text: http://www.brown.edu/Research/Primate/lpn30-1.html#loon

Pruetz JD, Bloomsmith MA 1992. Comparing two manipulable objects as enrichment for captive chimpanzees. Animal Welfare 1, 127-137
  FT: http://www.awionline.org/Lab_animals/biblio/aw1-12.htm

Ranheim S, Reinhardt V 1989. Compatible rhesus monkeys provide long-term stimulation for each other. Laboratory Primate Newsletter 28(3), 1-2
  FT: http://www.brown.edu/Research/Primate/lpn28-3.html#vik

Reese EP 1991. The role of husbandry in promoting the welfare of laboratory animals. In Animals in Biomedical Research Hendriksen CFM, Koeter HBWM (eds), 155-192. Elsevier, Amsterdam, Netherlands

Reinhardt V 1989a. Behavioral responses of unrelated adult male rhesus monkeys familiarized and paired for the purpose of environmental enrichment. American Journal of Primatology 17, 243-248
  FT: http://www.brown.edu/Research/Primate/lpn27-4.html#vik

Reinhardt V 1989b. Evaluation of the long-term effectiveness of two environmental enrichment objects for singly caged rhesus macaques. Lab Animal 18(6), 31-33
  FT: http://www.awionline.org/Lab_animals/biblio/la-eval.htm

Reinhardt V, Houser WD, Eisele S 1989. Pairing previously singly caged rhesus monkeys does not interfere with common research protocols. Laboratory Animal Science 39, 73-74

Reinhardt V 1990a. Time budget of caged rhesus monkeys exposed to a companion, a PVC perch and a piece of wood for an extended time. American Journal of Primatology 20, 51-56

Reinhardt V 1990b. Avoiding undue stress: Catching individual animals in groups of rhesus monkeys. Lab Animal 19(6), 52-53
  FT: http://www.awionline.org/Lab_animals/biblio/la-avoid.htm

Reinhardt V 1990c. Catching Individual Rhesus Monkeys Living in Captive Groups (Videotape with commentary). Wisconsin Regional Primate Research Center, Madison (Available on loan from Animal Care Audio-Visual Materials, WRPRC, 1220 Capitol Court, Madison, WI 53715, USA)

Reinhardt V, Cowley D 1990. Training stumptailed monkeys to cooperate during in-homecage treatment. Laboratory Primate Newsletter 29(4), 9-10
  FT: http://www.brown.edu/Research/Primate/lpn29-4.html#vik

Reinhardt V 1991. Training adult male rhesus monkeys to actively cooperate during in-homecage venipuncture. Animal Technology 42, 11-17
  FT: http://www.awionline.org/Lab_animals/biblio/at11.htm

Reinhardt V, Pape R 1991. An alternative method for primate perch installation. Lab Animal 20(8), 47-48
  FT: http://www.awionline.org/Lab_animals/biblio/la-an.htm

Reinhardt V, Reinhardt A 1991. Impact of a privacy panel on the behavior of caged female rhesus monkeys living in pairs. Journal of Experimental Animal Science 34, 55-58
  Full Text: http://www.awionline.org/Lab_animals/biblio/es34-5~1.htm

Reinhardt V, Cowley D, Eisele S 1991. Serum cortisol concentrations of single-housed and isosexually pair-housed adult rhesus macaques. Journal of Experimental Animal Science 34, 73-76
  Full Text: http://www.awionline.org/Lab_animals/biblio/es34-7~1.htm

Reinhardt V 1992a. Environmental enrichment branches that do not clog drains. Laboratory Primate Newsletter 31(2), 8
  Full Text: http://www.brown.edu/Research/Primate/lpn31-2.html#branch

Reinhardt V 1992b. Transport-cage training of caged rhesus macaques. Animal Technology 43, 57-61
  FT: http://www.awionline.org/Lab_animals/biblio/at57.htm

Reinhardt V 1992c. Improved handling of experimental rhesus monkeys. In The Inevitable Bond. Examining Scientist-Animal Interactions Davis H, Balfour AD (eds), 171-177. Cambridge University Press, Cambridge, UK
  FT: http://www.awionline.org/Lab_animals/biblio/bond.htm

Reinhardt V 1993a. Using the mesh ceiling as a food puzzle to encourage foraging behaviour in caged rhesus macaques (*Macaca mulatta*). Animal Welfare 2, 165-172
  Full Text: http://www.awionline.org/Lab_animals/biblio/aw3mesh.htm

Reinhardt V 1993b. Enticing nonhuman primates to forage for their standard biscuit ration. Zoo Biology 12, 307-312
  FT: http://www.awionline.org/Lab_animals/biblio/zb12-30.htm

Reinhardt V 1994a. Social enrichment for previously single-caged stumptail macaques. Animal Technology 5, 37-41
  FT: http://www.awionline.org/Lab_animals/biblio/at37.htm

Reinhardt V 1994b. Pair-housing rather than single-housing for laboratory rhesus macaques. Journal of Medical Primatology 23, 426-431
  FT: http://www.awionline.org/Lab_animals/biblio/jmp23.htm

Reinhardt V 1994c. Caged rhesus macaques voluntarily work for ordinary food. Primates 35, 95-98
  FT: http://www.awionline.org/Lab_animals/biblio/primat~1.htm

Reinhardt V, Liss C, Stevens C 1995. Restraint methods of laboratory nonhuman primates: A critical review. Animal Welfare 4, 221-238
  Full Text: http://www.awionline.org/Lab_animals/biblio/aw6metho.htm

Reinhardt V 1997. The Wisconsin Gnawing Stick. Animal Welfare Information Center (AWIC) Newsletter 7(3-4), 11-12
  Full Text: http://www.nal.usda.gov/awic/newsletters/v7n3/7n3reinh.htm

Reinhardt V 1999. Pair-housing overcomes self-biting behavior in macaques. Laboratory Primate Newsletter 38(1), 4
  Full Text: http://www.brown.edu/Research/Primate/lpn38-1.html#pair

Reinhardt V, Reinhardt A 1999. Are legal cage space requirements sound? Laboratory Primate Newsletter 38(2), 5-6
  Full Text: http://www.brown.edu/Research/Primate/lpn38-2.html#cage

Reinhardt V, Garza-Schmidt M 2000. Daily feeding enrichment for laboratory macaques: Inexpensive options. Laboratory Primate Newsletter 39(2), 8-10
  FT: http://www.brown.edu/Research/Primate/lpn39-2.html#vik

Riviello MC, Misiti A 1995. An alternative to woodchip as a foraging substrate for tufted capuchin monkeys (*Cebus apella*). Primate Report 42, 24
  Full Text: http://www.awionline.org/Lab_animals/biblio/pr42-2~1.htm

Schapiro SJ, Bushong D 1994. Effects of enrichment on veterinary treatment of laboratory rhesus macaques (*Macaca mulatta*). Animal Welfare 3, 25-36
FT: http://www.awionline.org/Lab_animals/biblio/aw3-25.htm

Schapiro SJ, Nehete PN, Perlman JE, Sastry KJ 1997. A change in housing condition leads to relatively long-term changes in cell-mediated immune responses in adult rhesus macaques. American Journal of Primatology 42, 146

Schapiro SJ, Nehete PN, Perlman JE, Sastry KJ 2000a. A comparison of cell-mediated immune responses in rhesus macaques housed singly, in pairs, or in groups. Applied Animal Behaviour Science 68, 67-84

Schapiro SJ, Stavisky R, Hook M 2000b. The lower-row cage may be dark, but behaviour does not appear to be affected. Laboratory Primate Newsletter 39(1), 4-6
Full Text: http://www.brown.edu/Research/Primate/lpn39-1.html#dark

Schnell CR, Gerber P 1997. Training and remote monitoring of cardiovascular parameters in non-human primates. Primate Report 49, 61-70
Full Text: http://www.awionline.org/Lab_animals/biblio/pr49-6~1.htm

Segerson L, Laule GE 1995. Initiating a training program with gorillas at the North Carolina Zoological Park. American Zoo and Aquarium Association (AZA) Annual Conference Proceedings, 488-489

Shideler SE, Savage A, Ortuño AM, Moorman EA, Lasley BL 1994. Monitoring female reproductive function by measurement of fecal estrogen and progesterone metabolites in the white-faced saki (*Pithecia pithecia*). American Journal of Primatology 32, 95-108

Shively CA 2001. Psychological Well-Being of Laboratory Primates at Oregon Regional Primate Research Center. Portland: Willamette Week, March 21, 2001
Full Text: http://www.wweek.com/html2/shivreport.html

Smith A, Lindburg DG, Vehrencamp S 1989. Effect of food preparation on feeding behavior of lion-tailed macaques. Zoo Biology 8, 57-65

Smith CC, Ansevin A 1957. Blood pressure of the normal rhesus monkey. Proceedings of the Society for Experimental Biology and Medicine 96, 428-432

Spragg SDS 1940. Morphine addiction in chimpanzees. Comparative Psychology Monographs 15, 1-132

Sokol KA 1993. Commentary: Thinking like a monkey—"primatomorphizing" an environmental enrichment program. Lab Animal 22(5), 40-45

Taylor WJ, Brown DA, Lucas-Awad J, Laudenslager ML 1997. Response to temporally distributed feeding schedules in a group of bonnet macaques (*Macaca radiata*). Laboratory Primate Newsletter 36(3), 1-3
Full Text: http://www.brown.edu/Research/Primate/lpn36-3.html#taylor

Turkkan JS 1990. New methodology for measuring blood pressure in awake baboons with use of behavioral training techniques. Journal of Medical Primatology 19, 455-466
FT: http://www.awionline.org/Lab_animals/biblio/jmp19-4.htm

Turkkan JS, Ator NA, Brady JV, Craven KA 1989. Beyond chronic catheterization in laboratory primates. In Housing, Care and Psychological Wellbeing of Captive and Laboratory Primates Segal EF (ed), 305-322. Noyes Publications, Park Ridge, NJ

United States Department of Agriculture 1991. Title 9, CFR (Code of Federal Register), Part 3. Animal Welfare; Standards; Final Rule. Federal Register 56(No. 32), 6426-6505
Full Text: http://www.nal.usda.gov/awic/legislat/awadog.htm

United States Department of Agriculture 2000. Animal Welfare Report—Fiscal Year 2000. U.S. Department of Agriculture—Animal Care, Riverdale, MD
Full Text: http://www.aphis.usda.gov/ac/awrep2000.pdf

de Waal FBM 1992. A social life for chimpanzees in captivity. In Chimpanzee Conservation and Public Health: Environments for the Future Erwin J, Landon JC (eds), 83-87. Diagnon/Bioqual, Rockville, MD

van Wagenen G 1950. The monkeys. In The Care and Breeding of Laboratory Animals Farris EJ (ed), 1-42. John Wiley, New York, NY

Wakenshaw V 1999. The management and husbandry of Geoffroy's marmoset. International Zoo News 46(1), 3-15
FT: http://www.awionline.org/Lab_animals/biblio/izn-wak.htm

Watson DSB 1991. A built-in perch for primate squeeze cages. Laboratory Animal Science 41, 378-379

Weed JL, Watson LM 1998. Pair housing adult owl monkeys (*Aotus* sp.) for environmental enrichment. American Journal of Primatology 45, 212

Weick BG, Perkins SE, Burnett DE, Rice TR, Staley EC 1991. Environmental enrichment objects and singly housed rhesus monkeys: Individual preferences and the restoration of novelty. Contemporary Topics in Laboratory Animal Science 30(5), 18

White G, Hill W, Speigel G, Valentine B, Weigant J, Wallis J 2000. Conversion of canine runs to group social housing for juvenile baboons. AALAS [American Association for Laboratory Animal Science] 51st National Meeting Official Program, 126

Williams LE, Abee CR, Barnes SR, Ricker RB 1988. Cage design and configuration for an arboreal species of primate. Laboratory Animal Science 38, 289-291

Wolfle TL 1987. Control of stress using non-drug approaches. Journal of the American Veterinary Medical Association 191, 1219-1221

Ziegler TE, Bridson WE, Snowdon CT, Eman S 1987. Urinary gonadotropin and estrogen excretion during the postpartum estrus, conception, and pregnancy in the cotton-top tamarin (*Saguinus oedipus oedipus*). American Journal of Primatology 12, 127-140

Viktor Reinhardt is Laboratory Animal Advisor to the Animal Welfare Institute in Washington, DC. He is a clinical veterinarian and ethologist and did extensive research in ethology and animal husbandry of nonhuman primates.

Comfortable Quarters for Pigs in Research Institutions

Temple Grandin
Department of Animal Science, Colorado State University, Fort Collins, CO 80523, USA

Under natural circumstances pigs live in small maternal groups of three to five sows with some juveniles (Frädrich, 1974). Gestating sows will temporarily leave the group, root a shallow hole in the ground and build a nest with branches and soft material, and give birth in seclusion (Stolba and Wood-Gush, 1989). Pigs spend most of their time foraging and eating. Rooting is a very important behavior, and individuals may show it with a frequency of about 60 times per 24 hours. A pig's snout and the rooting-disk is the universal tool for this animal. The rooting-disk consists of as many tactile receptors as a human hand and enables the pig to carefully explore the environment and search for food on and under the surface of the ground. Pigs are highly motivated to root, and will redirect this behavior to substitute objects—such as penmates—in a barren environment (van Putten, 1979).

Pigs are conspicuously sensitive animals who require special attention to guarantee their physical and behavioral well-being in the often stressful environment of a research institution. A stressed pig will yield stressed research data of little or no value, while a well kept and carefully handled pig is more likely to yield reliable research data of scientific value.

Figure 1. Author petting pigs. Contact with people produces calmer animals. The white cloth strips provide the pigs with opportunities for chewing and playing. The animals prefer soft pliable enrichment objects such as cloth strips to hard objects such as chains.

- **Provide pigs with roughage such as oat hulls and straw so that the animals can perform foraging activities and fill their gut.** Abnormal behaviors in pigs can be prevented or ameliorated by the provision of straw (van Putten, 1980; Arellano et al., 1992; Arey, 1993; Whittaker et al., 1998). Long natural straw is more effective than chopped straw. If experiments last more than a few weeks, avoid using pigs who have been bred for rapid weight gain and consequently have such a high feeding motivation that supplemental roughage may not satisfy them. Such animals are susceptible to developing stereotypical behavior and experiencing impaired well-being if they are kept on a calorie-restricted diet (Appleby and Lawrence, 1987; Terlouw et al., 1990).
- **Enrich the environment of pigs to counteract frustration and boredom resulting from chronic understimulation.** A barren concrete floor offers no stimulation, and it may cause discomfort and pain resulting from foot and joint irritations and inflammations. This is most likely to be a problem in lean hybrids. The author has observed that some breeding companies do not select for sound feet and legs. They have selected for rapid lean growth and have ignored other important traits such as structural soundness (Grandin, 1998). Keep the animals on straw bedding (cf., Figure 3) to reduce physical discomfort and the risk of leg injuries, to promote foraging and rooting behaviors and to stimulate exploratory and play behavior (cf., Lay et al., 2000). Evidence suggests that pigs kept in a barren environment use penmates as substitutes for substrates leading to harmful social behavior (Buré, 1981; Hughes and Duncan, 1988; Fraser et al., 1991; Burbidge et al., 1994). Some lean hybrids have a more excitable disposition and they are more likely to engage in ear sucking than fatter types of pigs when they are housed on an unbedded concrete floor. This has to be avoided by all means. If extreme cleanliness is required, provide the pigs

Comfortable Quarters for Laboratory Animals Reinhardt V, Reinhardt A (eds), 78-82. Animal Welfare Institute, Washington, DC 20007

with suspended tires and hanging rubber hoses to play with and to chew on. Such objects can easily be sanitized.

Pigs will avoid enrichment objects that are contaminated with manure. To solve this problem toys should be suspended over the pens with rope or twine. This keeps them clean and still will allow the animals to manipulate them almost as freely as though they were on the floor of the pens (Grandin, 1986). Chains are not very attractive enrichment gadgets for pigs (Horell and Ness, 1995; Hill et al., 1998). Pigs will play with chains, but they prefer to play with pliable objects when they are given a choice (Grandin, 1988). Cloth strips are particularly effective enrichment items (Grandin and Curtis, 1984). Strips of old bedsheet about 3 in. wide and 24 in. long can be easily tied to the fence and replaced when they get dirty (Figure 1). Environmental enrichment will make the animals calmer and less likely to be startled by sudden noise (Grandin, 1988; Grandin et al., 1987) or by people (Pearce et al., 1989; Moore et al., 1994). Pigs who are distracted in species-adequate ways will also show less apathy and engage in less stereotypical (Apple and Craig, 1992; Haskell et al., 1995) and anti-social activities (Schaefer et al., 1990; Arey, 1993) such as biting (Fraser, 1975; Beattie et al., 1995) and chewing ears and tails of their penmates (van Putten, 1979; Simonsen, 1990; Beattie et al., 1993).

- **Preparturient sows must have access to straw** (Figure 2). The importance of straw approaches that of feed, particularly during the 24 hours before farrowing when sows are highly motivated to build a nest for their young (Arey, 1992). The performance of nest-building behavior is in itself reinforcing independent of functional results (Hutson, 1992). An argument against straw in the research laboratory is the possibility of straw clogging drainage systems; a trap on the drain can readily avoid this problem (Batchelor, 1991).

- **Provide pigs with regular positive human contact in their home pens.** Attending care personnel should enter the pens every day and interact with the animals by allowing them to approach without fear and by stroking them firmly and talking to them gently (Figure 1). This will produce calmer animals who will be less fearful of humans (Tanida et al., 1995) and less excited and less stressed during research procedures (Grandin et al., 1983). The author has found that playing a radio at a reasonable volume helps to prevent excitement when pigs hear a sudden noise. Choose a station with a variety of talk and music. A radio with a lot of talk gets pigs accustomed to human voices.

- **House pigs in groups to account for their strong social disposition.** Keeping the animals in compatible groups is a safeguard for their behavioral health and general well-being (Figure 3). Avoid disrupting the composition of the group; this will help the animals to form stable social relationships and avoid aggressive conflicts. Provide the animals sufficient space and visual barriers to minimize aggressive conflicts arising from confinement. Housing with increased space allowance and partial stalls has welfare advantages in the long term, on the basis of decreased aggression, reduced cortisol concentrations

Figure 2. Preparturient sows have a strong instinct to build a nest for their piglets (photo by Diane Halverson, Animal Welfare Institute, Washington, DC).

and increased immunological responsiveness (Barnett et al., 1992; Waren and Broom, 1993; Andersen et al., 1999). A minimum space of between 2.4 and 3.6 m^2/adult pig is necessary to avoid undue, injurious aggression and hence promote good welfare (Weng et al., 1998).

Pigs form stable groups, but there is a tendency to fight if unfamiliar animals are placed together. Tailbiting and overt aggression are common results of mixing, particularly if done in confined spaces. To avert this, pigs should be mixed on neutral territory (Wolfenshohn and Lloyd, 1994). There are welfare advantages in reducing aggression by grouping unfamiliar adult pigs after dark and subsequently providing ad libitum food to reduce aggression at feeding (Barnett et al., 1994).

- **When a pig has to be removed from the home pen for a research procedure, always take another familiar pig along.** Pigs are social animals who quickly become apprehensive, agitated, and distressed when they are separated from their herdmates.

- **Do not house pigs alone unless there is a good veterinary reason or an IACUC-approved research reason for it.** If the research protocol requires temporary single-housing, allow the animal to keep visual contact with other familiar pigs to mitigate negative long-term effects of isolation (Herskin and Jensen, 2000). If this is not possible, provide the pig extra physical contact with

Figure 3. Pigs have a need for social companionship. Access to straw will help prevent behavioral problems (photo by Diane Halverson, Animal Welfare Institute, Washington, DC).

a friendly, familiar person. Patting a pig is not a waste of time because it improves scientific methodology by making the animal feel more at ease and increasing his or her coping capacity in a potentially distressing situation (Geers et al., 1995).

Provide single-housed animals enough room not only to lie fully relaxed without head or nose touching the feeder but also to turn around freely.

- **Allow the pigs to run up and down the aisle at least once a week.** This gives the animals some extra exercise and reduces their fear when they have to be moved from their pen to another location. Moving pigs in the aisle and walking through their pens every day will make the animals easier to handle (Abbott et al., 1997; Grandin, 2000), thereby minimizing or eliminating data-biasing stress reactions.

- **Do not subject pigs to forced movement and handling procedures.** Pigs are intelligent animals and—with a good understanding of their species-typical behavior and sensory physiology—can be readily trained to cooperate during common research procedures. When training pigs to work with rather than against you, never punish them but reward them with highly palatable treats. The training does take a little extra time, but researchers who have used trained pigs in their laboratories find that the benefits are worth the time investment. Trained animals experience less stress and hence provide more reliable research data. Needless to say, a well-designed training program also serves as a mental stimulation not only for the pigs but also for the care personnel.

- **Do not tether pigs.** Being tethered is a chronic stress situation for them (Barnett et al., 1985; Barnett et al., 1987; Schouten et al., 1991), which is likely to negatively influence research data.

- **Pigs should not be restrained unless it is absolutely necessary.** The animals must be completely tamed and acclimated to people before being trained to accept a restraint device. Slings make the best restraints for pigs. The pressure on the belly seems to have a calming effect. When sling-restrained, most Yucatan miniature swine will not only lie quietly but often sleep during the experimental procedures (Panepinto et al., 1983). Pressure on the side of a pig's body will cause the animal to relax (Grandin et al., 1989). The time period of restraint should always be the very minimum required to accomplish a specific research objective (National Research Council, 1996).

References

Abbott TA, Hunter EJ, Guise HJ, Penny RHC 1997. The effect of experience of handling on pigs' willingness to move. Applied Animal Behaviour Science 54, 371-375

Andersen IL, Bøe KE, Kristiansen AL 1999. The influence of different feeding arrangements and food type on competition at feeding in pregnant sows. Applied Animal Behaviour Science 65, 91-104

Apple JK, Craig JV 1992. The influence of pen size on toy preference of growing pigs. Applied Animal Behaviour Science 35, 149-155

Appleby MC, Lawrence AB 1987. Food restriction as a cause of stereotypic behaviour in tethered gilts. Animal Production 45, 103-110

Arellano PE, Pijoan C, Jacobson LD, Algers B 1992. Stereotyped behaviour, social interactions and suckling pattern of pigs housed in groups or in single crates. Applied Animal Behaviour Science 35, 157-166

Arey DS 1993. The effect of bedding on the behaviour and welfare of pigs. Animal Welfare 2, 235-246

Arey DS 1992. Straw and food as reinforcers for prepartal sows. Applied Animal Behaviour Science 33, 217-226

Barnett JL, Cronin GM, McCallum TH, Newman EA 1994. Effects of food and time of day on aggression when grouping unfamiliar adult pigs. Applied Animal Behaviour Science 39, 339-347

Barnett JL, Hemsworth PH, Cronin GM, Newman EA, McCallum TH, Chilton D 1992. Effects of pen size, partial stalls and method of feeding on welfare-related behavioural and physiological responses in group-housed pigs. Applied Animal Behaviour Science 34, 207-202

Barnett JL, Hemsworth PH, Winfield CG, Fahy VA 1987. The effects of pregnancy and parity number on behavioural and physiological responses related to the welfare status of individual and group-housed pigs. Applied Animal Behaviour Science 17, 229-243

Barnett JL, Winfield CG, Cronin GM, Hemsworth PH, Dewar AM 1985. The effect of individual and group housing on behavioural and physiological responses related to the welfare of pregnant pigs. Applied Animal Behaviour Science 14, 149-161

Batchelor GR 1991. Environment enrichment for the laboratory pig. Animal Technology 42, 185-189

Beattie VE, Sneddon IA, Walker N 1993. Behaviour and productivity of the domestic pig in barren and enriched environments. In Livestock Environment IV. Fourth International Symposium 42-50. American Society of Agricultural Engineers, St Joseph, MI

Beattie VE, Walker N, Sneddon IA 1995. Effect of rearing environment and change of environment on the behaviour of gilts. Applied Animal Behaviour Science 46, 57-65

Buré RG 1981. Animal well-being and housing systems for piglets. In The Welfare of Pigs Sybesma W (ed), 198-207. Martinus Nijhoff, The Hague, Netherlands

Burbidge JA, Spoolder HAM, Lawrence AB, Simmins PH, Edwards SA 1994. The effect of feeding regime and the provision of a foraging substrate on the development of behaviours in group-housed sows. Applied Animal Behaviour Science 40, 72

Frädrich H 1974. A comparison of the behaviour of the *Suidae*. In The Behaviour of Ungulates and its Relation to Management, IUCN New Series, No. 24 Geist V, Walther F (eds), 133-143. International Union for Conservation of Nature and Natural Resources [IUCN], Morges, Switzerland

Fraser D 1975. The effect of straw on the behaviour of sows in tethered stalls. Animal Production 21, 59-68

Fraser D, Phillips PA, Thompson BK, Tennessen T 1991. Effect of straw on the behaviour of growing pigs. Applied Animal Behaviour Science 30, 307-318

Geers R, Janssens G, Villé H, Bleus E, Gerard H, Janssens S, Jourquin J 1995. Effect of human contact on heart rate of pigs. Animal Welfare 4, 315-359

Grandin T 1988. Environmental enrichment for confinement pigs. Livestock Handling Committee Proceedings of the 1988 Annual Meeting, Kansas City, MO
Full Text: http://grandin.com/references/LCIhand.html

Grandin T 1986. Minimizing stress in pig handling. Lab Animal 15(3), 15-20

Grandin T 1998. Benefits and animal welfare. In Genetics and the Behavior of Domestic Animals Grandin T (ed), 319-346. Academic Press, San Diego, CA

Grandin T 2000. Handling and welfare of livestock in slaughter plants. In Livestock Handling and Transport Grandin T (ed), 409-439. CAB International, Wallingford, UK

Grandin T, Curtis SE 1984. Material affected cloth-toy touching and biting by pigs. Journal of Animal Science 59(Supplement 1), 50

Grandin T, Curtis SE, Greenough WT 1983. Effects of rearing environment on the behavior of young pigs. Journal of Animal Science 57(Supplement 1), 137
Full Text: http://grandin.com/references/abstract-11.html

Grandin T, Curtis SE, Taylor IA 1987. Toys, mingling and driving reduce excitability in pigs. Journal of Animal Science 65(Supplement 1), 230
Full Text: http://grandin.com/references/abstract-6.html

Grandin T, Dodman N, Shuster L 1989. Effect of naltrexone on relaxation influenced by flank pressure in pigs. Pharmacology, Biochemistry and Behavior 33, 839-842

Haskell M, Wemelsfelder F, Mendl M, Calvert S, Lawrence AB 1995. The effect of barren and enriched housing environments on the interactive behaviour of pigs. Applied Animal Behaviour Science 44, 263

Herskin MS, Jensen KH 2000. Effects of different degrees of social isolation on the behaviour of weaned piglets kept for experimental purposes. Animal Welfare 9, 237-249

Hill JD, McGlone JJ, Fullwood SD, Miller MF 1998. Environmental enrichment influences on pig behavior, performance and meat quality. Applied Animal Behaviour Science 57, 51-68

Horrell I, Ness PA 1995. Enrichment satisfying specific behavioural needs in early-weaned pigs. Applied Animal Behaviour Science 44, 264

Hughes BO, Duncan IJH 1988. The notion of ethological "need" models of motivation and animal welfare. Animal Behaviour 36, 1696-1707

Hutson GD 1992. A comparison of operant responding by farrowing sows for food and nest-building materials. Applied Animal Behaviour Science 34, 221-230

Lay D, Haussmann MF, Daniels MJ 2000. Hoop housing for feeder pigs offers a welfare-friendly environment compared to a nonbedded confinement system. Journal of Applied Animal Welfare Science 3(1), 33-48

Moore EA, Broom DM, Simmins PH 1994. Environmental enrichment in flatdeck accomodation for exploratory behaviour in early-weaned piglets. Applied Animal Behaviour Science 41, 277-278

National Research Council 1996. Guide for the Care and Use of Laboratory Animals, 7th Edition. National Academy Press, Washington, DC

Panepinto LM, Phillips RW, Norden S, Pryor PC, Cox R 1983. A comfortable minimum stress method of restraint for Yucatan miniature swine. Laboratory Animal Science 33, 95-97

Pearce GP, Paterson AM, Pearce AN 1989. The influence of pleasant and unpleasant handling and the provision of toys on the growth of male pigs. Applied Animal Behaviour Science 23, 27-37

Schaefer AL, Salomons MO, Tong AKW, Sather AP, Lepage P 1990. The effect of environmental enrichment on aggression in newly weaned pigs. Applied Animal Behaviour Science 27, 41-52

Schouten WGP, Rushen J, de Passille AM 1991. Heart rate changes in loose and tethered sows around feeding. Applied Animal Behaviour Science 30, 173-196

Simonsen HB 1990. Behaviour and distribution of fattening pigs in the multi-activity pen. Applied Animal Behaviour Science 27, 311-324

Stolba A, Wood-Gush DGM 1989. The behaviour of pigs in a semi-natural environment. Animal Production 48, 419-425

Tanida H, Miura A, Tanaka T, Yoshimoto T 1995. Behavioral response to humans in individually handled weanling pigs. Applied Animal Behaviour Science 42, 249-259

Terlouw EMC, Lawrence AB, Lindstrom-Nielsen B, Illius AW 1990. The effect of food level and housing on the development of stereotypic behaviour in tethered sows. Applied Animal Behaviour Science 26, 295

van Putten G 1979. Ever been close to a nosey pig? Applied Animal Behaviour Science 5, 298

van Putten G 1980. Objective observation on the behaviour of fattening pigs. Animal Regulation Studies 3, 104-118

Waren NK, Broom DM 1993. The influence of a barrier on the behaviour and growth of early-weaned piglets. Animal Production 56, 115-119

Weng RC, Edwards SA, English PR 1998. Behaviour, social interactions and lesion scores of group-housed sows in relation to floor space allowance. Applied Animal Behaviour Science 59, 307-316

Whittaker X, Spoolder HAM, Edwards SA, Lawrence AB, Corning S 1998. The influence of dietary fibre and the provision of straw on the development of stereotypic behaviour in food restricted pregnant sows. Applied Animal Behaviour Science 61, 89-102

Wolfenshohn S, Lloyd M 1994. Handbook of Laboratory Animal Management and Welfare. Oxford University Press, Oxford, UK

Temple Grandin wrote her widely recognized Ph.D. thesis on the effect of rearing environment and environmental enrichment on behavior and neural development in young pigs. She is Assistant Professor of Animal Science at Colorado State University. Her extraordinary sensitivity for the needs of animals has made her an internationally respected consultant to the livestock industry.

Comfortable Quarters for Sheep in Research Institutions

Viktor and Annie Reinhardt
Animal Welfare Institute, PO Box 3650, Washington, DC 20007, USA

Sheep are probably the most apprehensive animals used in research. They require excellent professional care—including regular monitoring of their behavior—to address stress reactions to the inherently fear-provoking environment of the laboratory. If these stress reactions are not ameliorated or buffered, any research with sheep is questionable because the results are likely to be influenced by an important, yet uncontrolled variable. In the United States approximately 24,000 sheep are used in research (United States Department of Agriculture, 2000).

- Gregariousness is the most outstanding characteristic of sheep. A sheep needs to be with other sheep in order to be in a state of well-being and normative physiology (Figure 1). Sheep will always try to maintain uninterrupted visual contact with at least one other sheep, and they will flock close together at any sign of danger. Individuals show a multitude of endocrine, hematological and biochemical alterations (Kilgour and de Langen, 1970; McNatty and Young, 1973; Bobek et al., 1986; Parrott et al., 1987; Minton et al., 1992; Apple et al., 1993; van Adrichem and Vogt, 1993; Carbajal and Orihuela, 2001), stereotypic behaviors (Done-Currie et al., 1984) and a marked increase in heart rate and respiration rate when isolated from other sheep (Syme and Elphick, 1982; Baldock and Sibly, 1986; Carbajal and Orihuela, 2001). When a ewe is separated from other sheep by a fence, she will stay calm as long as she can see her companions. However, if the fence is solid, the isolated subject will show dramatic stress reactions (Baldock and Sibly, 1990).
- Like cattle, sheep establish well-defined social relationships with all other group members, and they follow certain leading animals when moving from one location to another (Scott, 1945). The harmony within a flock is based on stable dominance-subordination relationships, which predetermine harmonious interactions between partners. Any change in the group composition—removals or introductions—will destabilize or disrupt these relationships. This will prompt agonistic conflicts leading to the establishment of new rank relationships.
- Confined sheep are prone to develop wool pulling behavior. Individuals pick with their mouths strands of wool from the

Figure 1. Sheep do best in the company of other sheep in naturally structured outdoor enclosures with pasture (©STS, photo by H.P. Haering).

fleece of other—typically subordinate—group members, who in time can become quite denuded. As in captive muskox calves (*Ovibos moschatus;* Reinhardt and Flood, 1983, Figure 11), rabbits (Maertens and DeGroote, 1994) and nonhuman primates (Elton, 1979; Reinhardt et al., 1986), this seemingly abnormal behavior may reflect difficulties to adjust to abnormal, i.e., species-inadequate, stressful husbandry conditions.
- When given the opportunity sheep will spend approximately 8 hours per day grazing (Figure 2; Kilgour and Dalton, 1994) and travel several kilometers (Arnold and Dudzinski, 1978). Although casual observation of well-habituated sheep in small stalls restricting almost all locomotion may not reveal any outward signs of distress, the confined subjects will show a distinct endocrine stress response reflected in increased adrenal activity (Bowers et al., 1993).
- Unlike cattle, sheep are very fearful by nature, and any changes in their surroundings make them apprehensive and ready to flee. Novelty is usually an acute stressor for

Comfortable Quarters for Laboratory Animals Reinhardt V, Reinhardt A (eds), 83-88. Animal Welfare Institute, Washington, DC 20007

Figure 2. The grass-covered paddock is a practicable, species-appropriate housing environment for sheep kept in research institutions (©STS, photo by S. Scherrer).

Figure 3. Social housing of sheep is a fundamental condition of sound scientific research methodology.

them, a circumstance that makes the research laboratory a particularly unsuited place to collect data that are not tainted by stress. The shape of the human "predator" elicits a specially strong fear response that can only be overcome through patient and compassionate conditioning. Sheep have excellent memory of aversive experiences (Belschner, 1962). The experience of rough handling can make a sheep quasi "useless" for any subsequent experimental procedures.

Probably even more than cattle—who, unlike sheep can easily overcome their fear of people—**sheep require social company** in order to cope with the circumstances of their domestication. It therefore follows that, in accommodating these animals for purposes of experimentation, their social needs must be given full consideration, on a par with other well-recognized physical needs such as nutrition, hygiene, and shelter (Fraser, 1995; Figure 3). Data collected from a single-housed sheep are not representative of "normal" sheep, that is to say animals in the sheep-characteristic social setting (Bowers et al., 1993).

Stable social relationships provide the basic condition that the members of a flock will live in a relatively stress-free environment. To remove a sheep or to introduce a strange sheep into a flock is not a good idea because it will destabilize or disrupt dominance-subordination relationships leading to rank determining aggressive conflicts. Such conflicts are counterproductive in the research setting and should therefore be avoided by all means. Lambs should always be allowed to stay with their mother until the age of natural weaning. Premature separation from the ewe is extremely disturbing for the young who will exhibit behavioral and physiological responses indicative of stress, including increased levels of plasma cortisol and impaired immune responses (Price and Thos, 1980; Moberg and Wood, 1981; Napolitano et al., 1995).

Sheep do best in naturally structured outdoor enclosures with pasture (Figure 1). The grass-covered paddock is a practicable, species-appropriate housing environment for sheep kept in research institutions (Figure 2). The floor area per animal must be sufficient to permit adult animals to keep appropriate social distances and hence minimize aggressive conflicts arising from confinement, and to allow young animals to engage in play activities. Feed troughs should be shallow enough so that the animals can maintain visual contact with each other. There must be enough trough space and/or feed rack space so that all animals of the flock can feed at the same time.

Sheep do not adapt to the stress of separation from other sheep (Niezgoda et al., 1987). Individually housed sheep are restless and show an increased heart rate that may persist over many days and account for an increased metabolic rate of up to 15% (van Adrichem and Vogt, 1993). A lone sheep will never be at ease and hence will always be unsuited for research that relies on stress-free data (cf., Done-Currie et al., 1984; Marsden and Wood-Gush, 1986). Not only for scientific but also for ethical reasons, a sheep should not be housed alone unless there is evidence that social housing would jeopardize the animal's health and/or well-being. During experiments that require temporary single-housing—for example in metabolic stalls—provision must be made that the sheep can keep uninterrupted acoustic, visual, and olfactory contact while standing and lying with one or several other sheep kept in the same room in close proximity (cf., Olfert et al., 1993). The presence of mirror panels during isolation can reduce the magnitude but fails to eliminate the physiological responses to this psychologically distressing situation (Parrott et al., 1988). Since sheep treat their own reflection as a strange individual (Parrott, 1983), the mirror image may, actually, be mildly stressful (Franklin and Hutson, 1982). The adjustment to the stress associated with experimental conditions is best buffered by the presence of another familiar conspecific (Pearson and Mellor, 1976; Porter et al., 1995; cf., Lyons et al., 1988). This does not imply that the attending care personnel is unimportant; on the contrary.

Temporary experimental enclosures—such as stalls and crates—for individual animals should be no less than 2 x 1 m so that an adult sheep can freely turn around and make at least a few steps in one direction (Hinch and Lynch, 1997) and that a young sheep can engage in some rudimentary play activities.

A 15 cm layer of straw or coarse [to prevent "balling"] sawdust mixed with wood shavings are recommended as comfortable, species-appropriate **bedding** (Hinch and Lynch, 1997). When given the choice of floor surface, sheep show a clear preference for straw bedding compared with wooden slats (Gordon and Cockram, 1995; Figure 4a,b). Wooden slats are thought to assist with drying feces and

Figure 4a, b. Sheep prefer solid flooring with straw bedding (a; © Archiv STS) over wooden slat flooring (b; photo by Geoff N. Hinch).

keeping the sheep clean. However, they require regular maintenance to guarantee safe footing and to ensure that there are no irregular gaps and broken surfaces that could injure the animals.

The need to ruminate demands that sheep receive sufficiently **bulky food**. The overall fiber content of the ingested food should be no less than 40%. A lack of fiber may lead to the development of stereotypical bar chewing and wool pulling (Hinch and Lynch, 1997).

The characteristic **neophobia** of sheep makes it an imperative that care personnel, which includes not only animal caregivers and technicians but also the attending veterinarian(s) and the principal investigator(s), habituate all animals who are subjected to experimentation to any new, i.e., fear-inducing situations. Frequent gentle handling and familiarization are the key to habituate a ewe, ram, or lamb so that he or she will accept a potentially aversive situation, such as blood collection (Kilgour, 1987). It goes without saying that a person who does not get along well with animals in general, and with sheep in particular, and lacks understanding in the way sheep perceive their environment and the people in it should not be allowed to handle sheep in the research setting. Such a person will be a source of distress, affecting not only the well-being of the sheep but also the quality of research data collected.

With knowledge of sheep behavior, patience, and gentleness, the inherent **fear** of the human "predator" can easily be overcome and a relationship between sheep and human handler developed that is based on trust. Stroking sheep and quietly talking to them in the research laboratory setting has nothing to do with sentimentalism but with sound scientific methodology. Tame sheep have more normal cortisol concentrations and heart rates and hence are more reliable research subjects than fearful sheep (Pearson and Mellor, 1976; Hargreaves and Hutson, 1990). It is not an overstatement to say that good care personnel instill qualities in the animals that make them better and more reliable research subjects. Stress, on the other hand, leads to physiological and behavioral changes that increase the variability of the data and decrease the reliability of the results.

The **attending caretaker** or technician is at the pinnacle of a cascading series of environmental and social influences that determine the well-being of the animals. She or he must strive to develop a social bond with the sheep in her or his charge. This bond will convey to the individual sheep a quiet sense of assurance upon which coping strategies can be developed (Wolfle, 1996). Animal care personnel should be strongly encouraged to establish positive relationships with all the sheep in their charge. This implies that provision has to be made that a certain amount of time is set aside each day of the week, to allow attending personnel to interact with the animals in a manner that fosters a trustful relationship. This extra time investment quickly pays off in better research data, and in a more satisfactory, i.e., *humane* work environment (cf., Kidd, 1994).

Individual sheep will spontaneously lead others during routine handling procedures, such as weighing and veterinary examination (Hutson, 1993). It is good advice to encourage this leader-following behavior rather than interfere in this biologically natural behavior by coercing the animals to move in an unorganized, usually chaotic fashion. The preferable means for movement is to allow sheep to proceed as a group, away from the handler and follow the leader to the target location. The use of fear stimuli such as dogs or noise must be avoided by all means because they unnecessarily frighten the animals. Taking advantage of voluntary, organized movement minimizes stress responses during handling, thereby enhancing the reliability of research data collected (Hinch and Lynch, 1997).

Positive food reinforcement, such as barley, can help both the sheep and the handling personnel to get the job done more quickly (Hutson, 1985). The important thing to remember is, that low-stress or no-stress handling can only be achieved with sheep who are not forced to move, but who are so confident with the handling personnel and with the environment that they move voluntarily to a designated location and accept a certain procedure. It has been documented that sheep can easily be trained to voluntarily enter a tilt table and accept brief immobilization for a grain reward (Grandin, 1989). Restraint does not have to be a distressing, data-biasing procedure as demonstrated by "fourteen out of sixteen ewes [who] returned for one or more additional passes" (Grandin, 1989, p. 200). Positive reinforcement training techniques to assure voluntary cooperation during such common procedures as blood collection and injection have been described in detail in other medium-size ungulates, namely goats and antelopes (Grandin et al., 1995; Lager, 1998; Phillips et al., 1997). It is recommended that these techniques be applied for sheep. There is no reason to assume that they are not equally useful in this species.

Like all animals in research laboratories, sheep respond best to gentle-and-firm handling. They should be held securely and, if the procedure allows it, kept with all four feet firmly on the ground. Sheep have sensitive skin and, therefore, should not be held by the wool. If it is necessary, a sheep can be made to sit up on his or her hindquarters while the handler holds the forelegs and provides firm-and-gentle support to the head and back region with his or her legs and body. In this position examinations and treatments can be carried out under the condition that they take only a few minutes.

Summary

The management of sheep in research institutions must address the following minimum requirements in order to be species-adequate and conducive to reliable research:

- Housing in stable group(s);
- Spacious indoor enclosure(s) with straw bedding;
- Access to spacious outdoor enclosure(s) with pasture;
- Provision of bulky food;
- Gentle familiarization to any new situation;
- Knowledgeable, patient and gentle-and-firm animal care personnel;
- Working **with** rather than against the animals during procedures.

References

Apple JK, Minton JE, Parson KM, Unruh JA 1993. Influence of repeated restraint and isolation stress and electrolyte administration on pituitary-adrenal secretions, electrolytes, and other blood constituents of sheep. Journal of Animal Science 71, 71-77

Arnold GW, Dudzinski ML 1978. Ethology of free-ranging domestic animals. Elsevier, Amsterdam, Netherlands

Baldock NM, Sibly RM 1990. Effects of handling and transportation on the heart rate and behaviour of sheep. Applied Animal Behaviour Science 28, 15-39

Baldock NM, Sibly RM 1986. Effects of management procedures on heart rate in sheep. Applied Animal Behaviour Science 15, 191

Belschner HG 1962. Sheep Management and Diseases, 7th Edition. Angus and Robertson, Sydney, Australia

Bobek S, Niezgoda J, Pierzchala K, Litynski P, Sechman A 1986. Changes in circulating levels of iodothyronines, cortisol and endogenous thiocyanate in sheep during emotional stress caused by isolation of animals from the flock. Zentralblatt für Veterinärmedizin 33, 698-705

Bowers CL, Friend TH, Grissom KK, Lay DC 1993. Confinement of lambs (Ovis aries) in metabolism stalls increased adrenal function, thyroxine and motivation for movement. Applied Animal Behaviour Science 36, 149-158

Carbajal S, Orihuela A 2001. Minimal number of conspecifics needed to minimize the stress response of isolated mature ewes. Journal of Applied Animal Welfare Science 4, 249-255

Done-Currie JR, Hecker JF, Wodzicka-Tomaszewska, M 1984. Behaviour of sheep transferred from pasture to an animal house. Applied Animal Behaviour Science 12, 121-130

Elton RH 1979. Baboon behavior under crowded conditions. In Captivity and Behavior Erwin J, Maple T, Mitchell G (eds), 125-139. Van Nostrand Reinhold, New York, NY

Franklin JR, Hutson GD 1982. Experiments on attracting sheep to move along a laneway. III. Visual stimuli. Applied Animal Ethology [Applied Animal Behaviour Science] 8, 457-478

Fraser AF 1995. Sheep. In The Experimental Animal in Biomedical Research, Volume II—Care, Husbandry, and Well-Being Rollin BE, Kesel ML (eds), 87-118. CRC Press, Boca Raton, FL

Gordon GDH, Cockram MS 1995. A comparison of wooden slats and straw bedding on the behaviour of sheep. Animal Welfare 4, 131-134

Grandin T 1989. Voluntary acceptance of restraint by sheep. Applied Animal Behaviour Science 23, 257-261

Grandin T, Rooney MB, Phillips M, Cambre RC, Irlbeck NA, Graffam W 1995. Conditioning of nyala (Tragelaphus angasi) to blood sampling in a crate with positive reinforcement. Zoo Biology 14, 261-273

Hargreaves AL, Hutson GD 1990. The effect of gentling on heart rate, flight distance and aversion of sheep to a handling procedure. Applied Animal Behaviour Science 26, 243-252

Hinch GN, Lynch JJ 1997. Comfortable quarters for sheep and goats. In Comfortable Quarters for Laboratory Animals Reinhardt V (ed), 94-100. Animal Welfare Institute, Washington, DC
Full Text: http://www.awionline.org/pubs/cq/sheep.htm

Hutson GD 1993. Behavioral principles of sheep handling. In Livestock Handling and Transport Grandin T (ed), 127-146. CAB International, Wallingford, UK

Hutson GD 1980. The effect of previous experience on sheep movement through yards. Applied Animal Ethology [Applied Animal Behaviour Science] 6, 233-240

Hutson GD 1985. The influence of barley food rewards on sheep movement through a handling system. Applied Animal Ethology [Applied Animal Behaviour Science] 14, 263-273

Kidd R 1994. Put away your prod: herd stock with less stress by understanding how they think. The New Farm 16(5), 6-10 & 44

Kilgour R 1987. Learning and the training of farm animals. The Veterinary Clinics of North America 3, 269-284

Kilgour R, Dalton DC 1994. Livestock Behaviour. Westview Press, Boulder, CO

Kilgour R, de Langen H 1970. Stress in sheep resulting from management practices. Proceedings of the New Zealand Society of Animal Production 30, 65-76

Lager K 1998. Apparatus and technique for conditioning goats to repeated blood collection. Lab Animal 27(3), 38-42

Lyons DM, Price EO, Moberg GP 1988. Social modulation of pituitary-adrenal responsiveness and individual differences in behaviour of young domestic goats. Physiology and Behavior 43, 451-458

Maertens L, DeGroote G 1984. Influence of the number of fryer rabbits per cage on their performance. Journal of Applied Rabbit Research 7, 151-155

Marsden MD, Wood-Gush DGM 1986. A note on the behaviour of individually-penned sheep regarding their use for research purposes. Animal Production 42, 157-159

McNatty KP, Young A 1973. Diurnal changes of plasma cortisol levels in sheep adapting to a new environment. Journal of Endocrinology 56, 329-330

Minton JE, Coppinger TR, Reddy PG, Davis WC, Blecha F 1992. Repeated restraint and isolation stress alters adrenal and lymphocyte functions and some leukocyte differentation antigens in lambs. Journal of Animal Science 70, 1126-1132

Moberg GP, Wood VA 1981. Neonatal stress in lambs: behavioral and physiological responses. Developmental Psychobiology 14, 155-162

Napolitano F, Marino V, de Rosa G, Capparelli R, Bordi A 1995. Influence of artificial rearing on behavioral and immune response of lambs. Applied Animal Behaviour Science 45, 245-253

Niezgoda J, Wronska D, Pierzchala K, Bobek S, Kahl S 1987. Lack of adaptation to repeated emotional stress evoked by isolation of sheep. Zentralblatt für Veterinärmedizin 34, 734-739

Olfert ED, Cross BM, McWilliam AA 1993. Guide to the Care and Use of Experimental Animals, Volume 1, 2nd Edition. Canadian Council on Animal Care, Ottawa, Canada
Full Text: http://www.ccac.ca/guides/english/V1_93/chap/chiv.htm#4B2

Parrott RF 1983. A method for the quantification of butting activity in androgen-treated wethers. Applied Animal Ethology [Applied Animal Behaviour Science] 10, 319-324

Parrot RF, Thornton SN, Forsling ML, Delaney E 1987. Endocrine and behavioural factors affecting water balance in sheep subjected to isolation stress. Journal of Endocrinology 112, 305-310

Parrott RF, Houpt KA, Misson BH 1988. Modification of the responses of sheep to isolation stress by the use of mirror panels. Applied Animal Behaviour Science 19, 331-338

Pearson RA, Mellor DJ 1976. Some behavioral and physiological changes in pregnant goats and sheep during adaptation to laboratory conditions. Research in Veterinary Science 20, 215-217

Phillips M, Grandin T, Graffam W, Irlbeck NA, Cambre RC 1997. Crate conditioning of bongo (Tragelaphus eurycerus) for veterinary and husbandry procedures at the Denver Zoological Gardens. Zoo Biology 17, 25-32

Porter R, Nowak R, Orgeur P 1995. Influence of a conspecific agemate on distress bleating by lambs. Applied Animal Behaviour Science 45, 239-244

Price EG, Thos J 1980. Behavioral responses of short-term isolation in sheep and goats. Applied Animal Ethology [Applied Animal Behaviour Science] 6, 331-339

Reinhardt V, Flood PF 1983. Behavioural assessment in muskox calves. Behaviour 87, 1-21

Reinhardt V, Reinhardt A, Houser WD 1986. Hair pulling-and-eating in captive rhesus monkeys. Folia Primatologica 47, 158-164
FT: http://www.awionline.org/Lab_animals/biblio/pr14-1.htm

Scott JP 1945. Social Behavior, Organization and Leadership in a Small Flock of Domestic Sheep—Comparative Psychology Monographs, Volume 18, Serial no. 96. Williams & Wilkins, Baltimore, MD

Syme LA, Elphick GR 1982. Heart rate and behaviour of sheep in yards. Applied Animal Ethology [Applied Animal Behaviour Science] 9, 31-35

United States Department of Agriculture 2000. Animal Welfare Report—Fiscal Year 2000. U.S. Department of Agriculture—Animal Care, Riverdale, MD
Full Text: http://www.aphis.usda.gov/ac/awrep2000.pdf

van Adrichem PWM, Vogt JE 1993. The effect of isolation and separation on the metabolism of sheep. Livestock Production Science 33, 151-159

Wolfle TL 1996. How different species affect the relationship. In The Human/Research Animal Relationship Krulisch L, Mayer S, Simmonds RC (eds), 85-91. Scientists Center for Animal Welfare, Greenbelt, MD

Viktor Reinhardt is Laboratory Animal Advisor to the Animal Welfare Institute in Washington, DC. He is a clinical veterinarian and ethologist and was responsible for the husbandry of sheep assigned to research at the Institute of Animal Husbandry and Hygiene in Bonn, Germany.

Annie Reinhardt is a librarian and manages the databases of the Animal Welfare Institute on Alternative Farming and on Refinement and Environmental Enrichment for Laboratory Animals. Annie has participated in numerous ethological studies conducted in cattle, muskox and bison.

Comfortable Quarters for Cattle in Research Institutions

Viktor and Annie Reinhardt
Animal Welfare Institute, PO Box 3650, Washington, DC 20007, USA

In order to provide cattle with appropriate housing and handling conditions in the research setting, their specific needs for behavioral, emotional, and physical well-being must be met. Only then can the researcher vouch that he/she adhered to basic principles of scientific methodology and ethics in his/her investigations.

- **Cattle are herd animals.** In the wild, the individual cow, bull, or calf is dependent on the presence of other conspecifics for the detection and avoidance of predators. A lone animal has little chance of survival. The members of a herd are tightly integrated in a cohesive social structure. Partners develop enduring, affectionate (Reinhardt and Reinhardt, 1981a; Figure 1), and stable dominance-subordination relationships (Reinhardt, 1980) that make the inter-individual interactions predictable and hence provide a safeguard for a harmonious co-existence as a coherent social group (Reinhardt, 1980). In sharp contrast to intensive housing systems, aggressive conflicts among cattle are rare, and even the presence of long horns poses no risk in a natural, free-ranging husbandry system where the animals have enough room for respecting each other's dominance-related spatial privileges thereby avoiding antagonism (Reinhardt, 1980; Figure 2). In the course of a 24-month study of a free-ranging cattle herd consisting of 29 cows, one bull and their offspring, several thousand agonistic interactions were witnessed, but an animal was never injured during such encounters (Reinhardt, 1980). Dominance rank is determined by age, which in turn correlates with horn length. Old animals who are no longer the heaviest and strongest animals of the herd do not lose their high position in the social hierarchy. Their long horns are an effective social weapon that help them retain the respect of younger animals (Reinhardt, 1980). When animals are dehorned, the biological foundation of the hierarchical system is destabilized, leading to alarmingly high rates of overt aggression (Reinhardt and Reinhardt, 1975).

In the research context, **group-housing** is the only acceptable husbandry system for cattle. Social isolation is a serious stressor leading to increased heart rate and plasma cortisol concentration, vocalization, and behavioral signs of frustration and fear (Dantzer and Mormède, 1983;

Figure 1. Cattle develop friendly relationships with one another that are based on mutual preferences. These two cows, *Nanette* and *Gilla,* are not kin-related but preferred each other—over 27 other cows of the herd—as grooming partner and as grazing partner (cf., Figure 5) during a test period of five years (Reinhardt and Reinhardt, 1981a).

Hopster and Blokhuis, 1994; Munksgaard and Simonsen, 1996; Veissier et al., 1997). Data collected under such circumstances are confounded by stress and, therefore, have little or no scientific value. The intensive need for a social partner can only be met by another cow. Contact with a familiar, empathetic person can ameliorate behavioral stress reactions, but it is not sufficiently comforting to reduce the endocrine response to isolation stress (Rushen et al., 2001).

It is imperative that cattle kept in a barn have as little reason as possible for competition. There must be enough **feeding space** and a sufficiently large **bedded area** so that <u>all</u> animals can comfortably access the food at the same time or rest in a comfortable recumbent position at the same time. This will foster group harmony (cf., Nielsen et al., 1997). Visual barriers help to buffer agonistic conflicts arising from the inherent spatial constraints of indoor

Comfortable Quarters for Laboratory Animals Reinhardt V, Reinhardt A (eds), 89-95. Animal Welfare Institute, Washington, DC 20007

Figure 2. Cattle do not use their horns to injure each other. Subordinate animals avoid overt aggression by moving out of the way of dominant partners. Note the bossy look of the cow at left and the moving-away gesture of the subordinate cow at right (drawing by Annie Reinhardt).

Figure 3. A strange cow [animal at left] submissively approaches another herd and is blocked from proceeding by a cow who displays the broadside-threat gesture (drawing by Ingrid Schaumburg).

Figure 4. Cow *Aida* nursing her newborn calf while grooming her 13-month old daughter whom she had weaned four months ago. *Aida's* mean calving interval was 338 days; she produced 9 calves—who were all allowed to stay with the maternal herd beyond the age of natural weaning— during a test period of 7.5 years (Reinhardt, 1983a; drawing by Ingrid Schaumburg).

housing (Bouissou, 1970). Sufficient space is necessary so that subordinate animals can yield to dominant partners, thus triggering no overt aggression (cf., Figure 2).

Stable rank relationships are a prerequisite so that there are no undue social tensions or overt conflicts. A group's composition should, therefore, not be altered unless there is a specific veterinary reason. The members of a cattle herd know each other intimately and will show xenophobic behavior towards other cattle (Reinhardt, 1980; cf., Schloeth, 1961; Scheurmann, 1975; Figure 3). To introduce a strange cow into a herd is not a good idea!

A loose-housing system that is designed to meet the animals' social spacing requirements along with management that respects the animals' hierarchy system and herd-feeling makes the **dehorning** of cattle unnecessary (Menke et al., 1999). Dehorning is a sign of inadequate husbandry, but it is also a distressing and painful experience for the animals (Taschke, 1995).

The **tie-stall** is an extremely uncomfortable [hard surface], painful [risk of inflammations of knees and hocks], frustrating [lying down is aversive, but there is a strong urge to rest in recumbence], and boring [restricted or no opportunity for social contact/interaction and foraging] housing environment (Krohn and Munksgaard, 1993; Redbo, 1993; Krohn, 1994; Haley et al., 2000) and, therefore, is not appropriate for cattle who are expected to yield research data that are not confounded by impaired well-being. The inadequacy of the tie-stall is reflected in the frequent occurrence of stereotypical activities [e.g., bar-biting, tongue-rolling], which disappear when the animals are transferred to loose housing or pasture (Redbo, 1992; Krohn, 1994). If circumstances require that a cow is temporarily tethered—e.g., venipuncture, remote sample collection via indwelling catheter—she should be tied by a halter and released as soon as the procedure is completed. A temporarily tied or single-housed animal must always be able to keep at least visual contact with other close-by members of the herd to buffer stress reactions. Under exceptional experimental circumstances lasting less than a day, a mirror may substitute for another conspecific (Piller et al., 1999).

- **Long-lasting affiliative relationships** exist not only among friends—who prefer each other as grazing, grooming, and resting partners (Reinhardt, 1980; Reinhardt and Reinhardt, 1981a; Reinhardt et al., 1986)—but also between mother and offspring. The bond between cow and calf is not affected by the weaning process—which occurs when the calf is approximately 10 months old (Reinhardt and Reinhardt, 1981b)—but lasts many years, leading to the development of tight-knit family subgroups within the herd (Reinhardt and Reinhardt, 1981a; Reinhardt et al., 1986; Figure 4).

In the research setting, distress has to be avoided to guarantee the scientific validity of research data. It is, therefore, not justifiable to subject the dam and the calf to the extremely disturbing situation created by premature **weaning.** Forced weaning distresses both the cow, who will show reduced reproductivity as a result of it (Reinhardt, 1982), and the calf, who will be prone to develop behavioral signs of frustration (Seo et al., 1998) and whose physiological ability to cope with stress will be

Figure 5. The grass field is the most appropriate living environment for cattle. Here two friends, *Nanette* and *Gilla*, keeping each other's company while grazing (cf., Figure 1).

Figure 6. Cattle must have access to shaded areas to forestall heat stress.

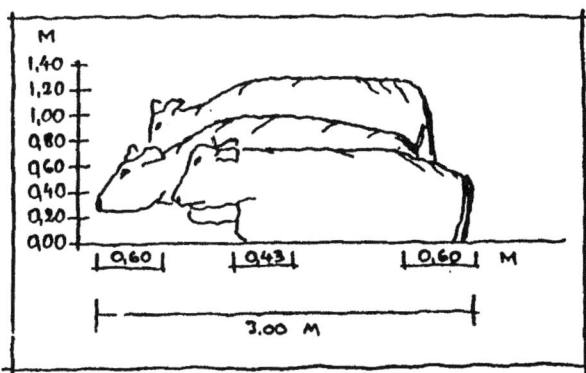

Figure 7. Cattle need considerable space in front of them to "swing" up into a standing position (drawing by H. Hoffmann).

impaired (Lay et al., 1992). If calves are allowed to stay with their mothers until the natural weaning process has occurred, they will not engage in compulsory substitute sucking, which often leads to health problems associated with the development of bezoars. It is not the substitute sucking that deserves the label "abnormal" (Loberg and Lidfors, 2001) but the human interference with a natural process (Reinhardt and Reinhardt, 1980a).

- Under natural conditions **cattle divide the day time into long periods of foraging** while moving considerable distances and long periods of chewing the cud while loafing or resting in a recumbent position (Reinhardt, 1980; Krohn and Munksgaard, 1997). When given the choice, cattle will spend almost all their time on a grass field rather than in a stable with deep bedding (Krohn et al., 1992). The urge to forage is so strong that well-fed cattle will push their way through a fenced area to get access to a meadow where they can graze (Trantham, 2000).

For cattle used in research, a well managed **pasture** is the most appropriate living environment (Krohn and Munksgaard, 1997; Figure 5). That's where they can graze ad libitum and that's where they can find suitable places to lie down comfortably and rest undisturbed in cattle-specific recumbent positions. The enhanced well-being of cattle on pasture is reflected by a high degree of herd synchrony and the absence of restlessness that typically occurs indoors as a result of spatial restriction (cf., O'Connell et al., 1989; Miller and Wood-Gush, 1991). Rotational grazing is the most suitable management system providing the animals adequate foraging opportunities, while fostering their health (Beetz, 1999).

Cattle seek out **shady places** during the hottest time of the day because they are very susceptible to heat stress (Kidd, 1993; Silanikove, 2000; Mitlöhner et al., 2001; Figure 6). If the pasture does not include trees or other shade-casting structures, a shademobile should be

Figure 8. Straw is a cattle-appropriate resting substrate and should be used whenever possible (©STS, photo by Hans-Peter Haering).

provided to ensure that the animals can avoid extreme exposure to direct sunlight during hot weather conditions (Salatin, 1991).

- **Cattle don't like to lie on hard or wet surfaces** but prefer dry, relatively soft locations for resting (Irps, 1983; Jensen et al., 1988). It should be remembered that recumbent animals need considerable space in front of them to "swing" up into a standing position (Figure 7; Hoffmann and Rist, 1975).

Under indoor-housing conditions, cattle should be provided with a dry, well heat-insulated, **straw-bedded lying area**. They prefer straw over sawdust, mats or slats as floor type (Lowe et al., 2001). Straw not only provides comfort while lying but also resistance to slipping and opportunity to express natural foraging behavior. Solid concrete and slatted floors, especially when they are slippery, dirty, and wet, create a serious risk of injuries (Anonymous, 2001). All animals must have free access to resting sites that are spacious enough to allow for unrestricted, cattle-specific resting postures and freedom of movement during postural changes. Straw is a cattle-appropriate resting substrate and should be used whenever possible (Krohn and Munksgaard, 1997; Figure 8). Lying mats can be equivalent to straw bedding in terms of cattle resting behavior, but they are less favorable with respect to leg injuries (Wechsler et al., 2000).

- **Cattle are herbivores.** They need rather voluminous fodder with a high fiber content for normal digestion and normal metabolism.

In the research setting, special care must be taken

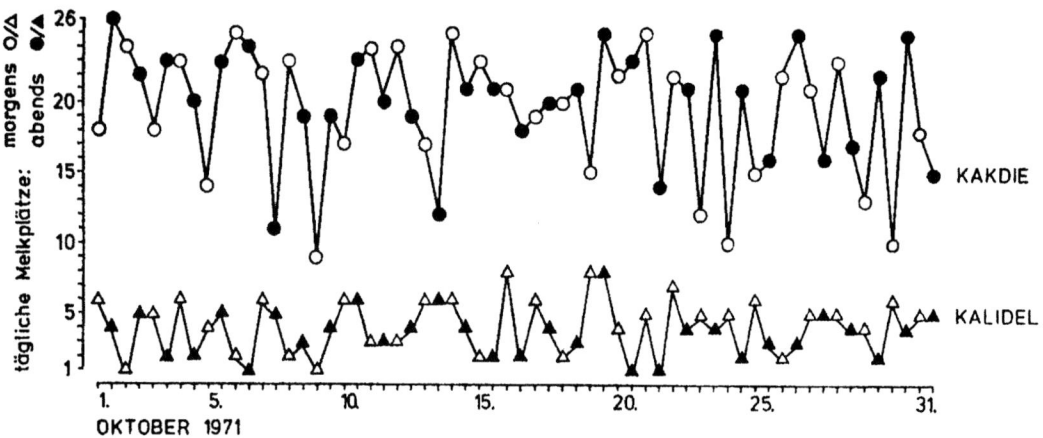

Figure 9. Positions in the order of entrance into the milking parlor of two cows—of a 26-animal herd—during the morning and afternoon milking on 31 consecutive days. The order of entrance is correlated with the social hierarchy; cow *Kalidel* had the social rank place No. 3, cow *Kakdie* place No. 21 (Reinhardt and Reinhardt, 1980b).

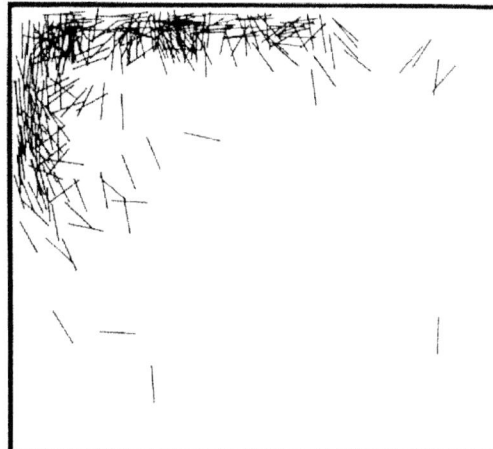 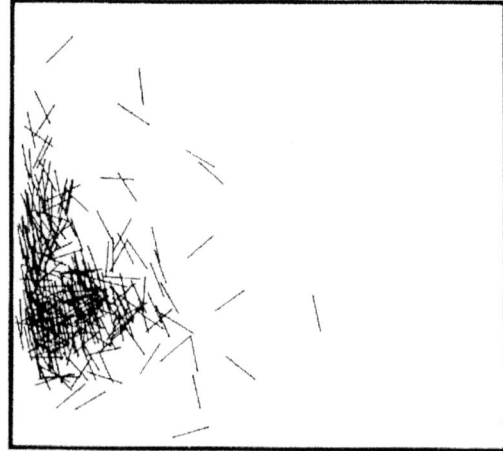

Figure 10. Resting positions of the two cows *Elsa* (left figure) and *Alma* (right figure)—members of a 29-cow herd [plus calves and one bull]—30 minutes after darkness in a 20 x 22 m resting area on 318 consecutive days (Reinhardt et al., 1978).

that cattle receive enough **roughage** to stimulate feeding and ruminating (Graf, 1994). A diet that is too concentrated may lead to metabolic imbalances, indigestion, and the development of behavioral disorders such as tongue rolling and nibbling at objects. Cattle who have ad libitum access to hay or regular access to a pasture do not develop these stereotypies.

- **Cattle develop strong habits in their daily routine.** When moving from one location to another—e.g., from the barn to the pasture—or when entering the milking parlor or the chute, they proceed in a predictable order in which each herd member has a well-defined place (Reinhardt, 1973; Reinhardt, 1983b; Andrade et al., 2001; Figure 9). When lying down, individual animals occupy specific resting sites with consistent preference (Schmisseur et al., 1966; Thinès et al., 1975; Reinhardt et al., 1978; Figure 10). These habits introduce a predictable routine in a herd's daily activities that should be respected by personnel. A cow may enter the milking parlor calmly when it is her turn but get obnoxious and stressed when coerced to do so too early. Interference in these habits also has a socially disruptive effect that disturbs not only one selected animal but the whole herd.

- **Cows, calves, but also bulls readily establish a positive relationship with humans.** They seem to enjoy being scratched by a trustworthy person no less than being groomed by another conspecific (Figure 11; cf., Norton et al., 2000). At the same time, however, cattle get easily frightened and stressed by a person who is hectic, impatient, or even callous (Kidd, 1994). Research data collected under such circumstances are confounded by stress and hence have no scientific value.

Animal caretakers and animal technicians must be knowledgeable of cattle behavior and cattle psychology (Grandin, 1989; Albright and Arave, 1997), and they must be willing and compassionate enough to establish a good relationship with the animals in their charge. Cattle who are well socialized with people are less fearful,

Figure 11. Cattle enjoy being groomed. Such a friendly interaction fosters a human-animal relationship that is based on trust rather than fear.

grow more quickly, have higher rates of milk production, and show improved immune competence (Mench et al., 1998). Professional, i.e., considerate handling implies that the research subject shows no signs of apprehension when being approached and no signs of fear, resistance or distress when being held in a stationary position, manipulated [e.g., for blood collection] or moved [e.g., enter an experimental stall]. This type of interaction increases the safety of the handling person, since the animal has no reason to show defense aggression; at the same time it buffers stress reactions, thereby preventing physiological changes that could confound research results. Cattle who have a good relationship with the

Figure 12a,b. A young cow warns you with a broadside-threat not to approach her (Figure 12a). If she attacks you (Figure 12b), there is no reason to blame her for a natural behavior reaction that is triggered by your inappropriate behavior. Good knowledge of cattle psychology is one of the basic prerequisites of good stockmanship.

handling personnel will show reliable baseline values of stress-sensitive parameters during procedures (Karg et al., 1972; Reinhardt and Schams, 1974). A positive human-animal relationship and good knowledge of bovine behavior also make the dehorning of cattle superfluous (Menke et al., 1999).

Cattle are not dumb beasts. They are very sensitive and quickly learn whom they can trust and whom they should avoid. They are not stubborn or aggressive by nature, but a brutal or unskilled person can easily trigger these reactions in them. In their behavioral responses, cattle—like all other confined animals—mirror the attitude of the people whom they are in contact with. When they exhibit so-called undesirable behavior, such as defensive aggression, it is always good advice to ask yourself "What is wrong with the way I treat them?" To blame animals for their reactions to our inappropriate behavior would reflect a high degree of ethological ignorance (Figure 12).

Recommendations

The following provisions are basic requirements for cattle-adequate husbandry in research institutions:

1. Social housing;
2. Sufficient feeding space and straw-bedded lying area for all group members;
3. Stable social relationships;
4. Access to pasture;
5. Positive human-animal relationship;
6. Firm-and-gentle handling.

References

Albright JL, Arave CW 1997. The Behaviour of Cattle. CabInternational, London, UK

Andrade O, Orihuela A, Solano J, Galina CS 2001. Some effects of repeated handling and the use of a mask on stress responses in zebu cattle during restraint. Applied Animal Behaviour Science 71, 175-181

Anonymous 2001. Scientists' assessment of the impact of housing and management on animal welfare. Journal of Applied Animal Welfare 4, 3-52

Beetz A 1999. Rotational grazing. ATTRA (Appropriate Technology Transfer for Rural Areas), Fayetteville, AR
Full Text: http://www.attra.org/attra-pub/rotategr.html

Bouissou MF 1970. Rôle du contact physique dans la manifestation des relations hiérarchiques chez les bovins: conséquences pratiques. Annales Zootechniques 19, 279-285

Dantzer R, Morméde P 1983. Physiological assessment of adaptation in farm animals. In Farm Animal Housing and Welfare Baxter SH, Baxter MR, MacCormack JAD (eds), 8-19. Marinus Nijhoff Publishers, Boston, MA

Graf B 1994. Abnormal oral behaviours in fattening bulls: incidence, causation and implications. Applied Animal Behaviour Science 40, 79-80

Grandin T 1989. Behavioral principles of livestock handling. Professional Animal Scientist 5(2), 1-11
Full Text: http://grandin.com/references/new.corral.html

Haley DB, Rushen J, de Passillé AM 2000. Behavioural indicators of cow comfort: activity and resting behaviour of dairy cows in two types of housing. Canadian Journal of Animal Science 80, 257-263

Hoffmann H, Rist M 1975. Tiergerecht und arbeitswirtschaftlich günstige Anbindevorrichtungen für Kühe. Schweizer Landwirtschaftliche Monatshefte 53, 119-126

Hopster H, Blokhuis HJ 1994. Consistent individual stress responses of dairy cows during social isolation. Applied Animal Behaviour Science 40, 83-84

Irps H 1983. Results of research projects into flooring preferences of cattle. In Farm Animal Housing and Welfare Baxter SH, Baxter MR, MacCormack JAC (eds), 200-215. Marinus Nijhoff, The Hague, Netherlands

Jensen MB, Vestergaard KS, Krohn CC 1998. Play behaviour in dairy calves kept in pens: the effect of social contact and space allowance. Applied Animal Behaviour Science 56, 97-108

Karg H, Schams D, Reinhardt V 1972. Effects of 2-Br-alpha-ergocryptine on plasma prolactin level and milk yield in cows. Experientia 28, 574-576

Kidd R 1993. Help livestock keep their cool: Water and shade are keys to comfort. The New Farm 15(5), 8-12
Full Text: http://www.awionline.org/farm/kidd2cool.html

Kidd R 1994. Put away your prod: herd stock with less stress by understanding how they think. The New Farm 16(5), 6-10 & 44
Full Text: http://www.awionline.org/farm/kidd1prod.html

Krohn CC 1994. Behaviour of dairy cows kept in extensive (loose housing/pasture) or intensive (tie stall) environments. III. Grooming, exploration and abnormal behaviour. Applied Animal Behaviour Science 42, 73-86

Krohn CC, Munksgaard L 1993. Behaviour of dairy cows kept in extensive (loose housing/pasture) or intensive (tie stall) environments II. Lying and lying-down behaviour. Applied Animal Behaviour Science 37, 1-16

Krohn CC, Munksgaard L 1997. Comfortable housing for cattle used in research. In Comfortable Quarters for Laboratory Animals Reinhardt V (ed), 101-106. Animal Welfare Institute, Washington, DC
Full Text: http://www.awionline.org/pubs/cq/cows.htm

Krohn CC, Munksgaard L, Jonasen B 1992. Behaviour of dairy cows kept in extensive (loose housing/pasture) or intensive (tie stall) environments. I. Experimental procedure, facilities, time budgets—diurnal and seasonal conditions. Applied Animal Behaviour Science 34, 37-47

Lay DC, Friend TH, Randel RD, Bowers CL, Neuendorff DA, Grissom KK, Jenkins OC 1992. Does maternal deprivation affect a calf's physiological and behavioral reactions to later stress? Journal of Animal Science 70 (Supplement 1), 162

Loberg J, Lidfors L 2001. Effect of milkflow rate and presence of a floating nipple on abnormal sucking between dairy calves. Applied Animal Behaviour Science 71, 189-199

Lowe DE, Steen RWJ, Beattie VE 2001. Preferences of housed finishing beef cattle for different floor types. Animal Welfare 10, 395-404

Mench JA, Morrow-Tesch J, Chu L 1998. Environmental enrichment for farm animals. Lab Animal 27(3), 32-36

Menke C, Waiblinger S, Fölsch DW, Wiepkema PR 1999. Social behaviour and injuries of horned cows in loose housing systems. Animal Welfare 8, 243-258

Miller K, Wood-Gush DGM 1991. Some effects of housing on the social behaviour of dairy cows. Animal Production 53, 271-278

Mitlöhner FM, Morrow JL, Dailey JW, Wilson SC, Galyean ML, Miller MF, McGlone JJ 2001. Shade and water misting effects on behavior, physiology, performance, and carcass traits of heat-stressed feedlot cattle. Journal of Animal Science 79, 2327-2335

Munksgaard L, Simonsen HB 1996. Behavioral and pituitary adrenal-axis responses of dairy cows to social isolation and deprivation of lying down. Journal of Animal Science 74, 769-778

Nielsen LH, Mogensen L, Krohn C, Hindhede J, Sorensen JT 1997. Resting and social behaviour of dairy heifers housed in slatted floor pens with different sized bedded lying areas. Applied Animal Behaviour Science 54, 307-316

Norton KC, Watach MJ, Gordon L, Litwak KN, Litwak P 2000. Environmental enrichment for calves with artificial organs. AALAS [American Association for Laboratory Animal Science] 51st National Meeting Official Program, 88

O'Connell J, Giller PS, Meaney W 1989. A comparison of dairy cattle behavioural patterns at pasture and during confinement. Irish Journal of Agricultural Research 28, 65-72

Piller CAK, Stookey JM, Watts JM 1999. Effects of mirror-image exposure on heart rate and movement of isolated heifers. Applied Animal Behaviour Science 63, 93-102

Redbo I 1992. The influence of restraint on the occurrence of oral sterotypies in dairy cows. Applied Animal Behaviour Science 35, 115-123

Redbo I 1993. Stereotypies and cortisol secretion in heifers subjected to tethering. Applied Animal Behaviour Science 38, 213-225

Reinhardt V 1973. Beiträge zur sozialen Rangordnung und Melkordnung bei Kühen. Zeitschrift für Tierpsychologie [Ethology] 32, 281-292

Reinhardt V, Schams D 1974. Analysis of teat stimulation as specific stimulus for prolactin in cattle. Neuroendocrinology 14, 289-296

Reinhardt V, Reinhardt A 1975. Dynamics of social hierarchy in a dairy herd. Zeitschrift für Tierpsychologie [Ethology] 38, 315-323

Reinhardt V, Mutiso FM, Reinhardt A 1978. Resting habits of zebu cattle in a nocturnal enclosure. Applied Animal Ethology [Applied Animal Behaviour Science] 4, 261-271

Reinhardt V 1980. Untersuchung zum Sozialverhalten des Rindes—Eine zweijährige Beobachtung an einer halb-wilden Rinderherde (Bos indicus). Birkhäuser Verlag, Boston, MA

Reinhardt V, Reinhardt A 1980a. Natürliche Säugegewohnheiten bei Kälbern. Landwirtschaftliche Zeitschrift 147, 866-867

Reinhardt V, Reinhardt A 1980b. Warum Kalidel immer zuerst gemolken wird. Landwirtschaftliche Zeitschrift 147, 1294 & 1299

Reinhardt V, Reinhardt A 1981a. Cohesive relationships in a cattle herd (Bos indicus). Behaviour 77, 121-151

Reinhardt V, Reinhardt A 1981b. Natural sucking performance and age of weaning in zebu cattle (Bos indicus). Journal of Agricultural Science 96, 309-312

Reinhardt V 1982. Reproductive performance in a semi-wild cattle herd (Bos indicus). Journal of Agricultural Science 98, 567-569

Reinhardt V 1983a. Reproduktionsdaten einer halb-wilden Rinderherde. Verhandlungsberichte der 7. Veterinärmedizinischen Gemeinschaftstagung, 124-126

Reinhardt V 1983b. Movement orders and leadership in a semi-wild cattle herd. Behaviour 83, 251-264

Rushen J, Munksgaard L, Marnet PG, DePassillé AM 2001. Human contact and the effects of acute stress on cows at milking. Applied Animal Behaviour Science 73, 1-14

Salatin J 1991. Profit by appointment only: This farm family puts quality first, and their customers love it. The New Farm 13(6), 8-12
Full Text: http://www.awionline.org/farm/salatin.html

Scheurmann E 1975. Observations of the behaviour of the Mithan (Bibos frontalis Lambert 1837) in captivity. Applied Animal Ethology 1, 321-355

Schloeth R 1961. Das Sozialverhalten des Camargue-Rindes. Zeitschrift für Tierpsychologie 18, 574-627

Schmisseur WE, Albright JL, Dillon WM, Kehrberg EW, Morris WHM 1966. Animal behaviour responses to loose and free stall housing. Journal of Dairy Science 49, 102-104

Seo T, Sato S, Kosaka K, Sakamoto N, Tokumoto K, Katoh K 1998. Development of tongue-playing in artificially reared calves: effects of offering a dummy-teat, feeding of short cut hay and housing system. Applied Animal Behaviour Science 56, 1-12

Silanikove N 2000. Effects of heat stress on the welfare of extensively managed domestic ruminants. Livestock Production Science 67(1-18)

Trantham T 2000. Twelve Aprils Dairying (Web site). FT: http://www.griffin.peachnet.edu/sare/twelve/trantham.html

Taschke A 1995. Ethologische, physiologische und histologische Untersuchungen zur Schmerzbelastung der Rinder bei der Enthornung. University of Zürich, Zürich, Switzerland

Thinès G, Soffié M, de Marneffe G 1975. Aires de résidence préférentielles d'un groupe de vaches laitières en stabulation libre. Annales Zootechniques 24, 177-187

Veissier I, Chazal P, Pradel P, Le Neindre P 1997. Providing social contacts and objects for nibbling moderates reactivity and oral behaviors in veal calves. Journal of Animal Science 75, 356-365

Wechsler B, Schaub J, Friedli K, Hauser R 2000. Behaviour and leg injuries in dairy cows kept in cubicle systems with straw bedding or soft lying mats. Applied Animal Behaviour Science 69, 187-197

Viktor Reinhardt is Laboratory Animal Advisor to the Animal Welfare Institute in Washington, DC. He is a clinical veterinarian and ethologist and did extensive research in reproductive physiology, animal husbandry and ethology in cattle, muskox and bison.

Annie Reinhardt is a librarian and manages the databases of the Animal Welfare Institute on Alternative Farming and on Refinement and Environmental Enrichment for Laboratory Animals. Annie has participated in numerous ethological studies conducted in cattle, muskox and bison.

Comfortable Quarters for Horses in Research Institutions

Katherine A. Houpt and T.S. Ogilvie-Graham

Houpt: Department of Biomedical Sciences, College of Veterinary Medicine, Cornell University, Ithaca, NY 14853-6401, USA

Ogilvie-Graham: Surgeon General's Department, St. Giles Court, London WC2H 8LD, United Kingdom

To address the species-typical behavior of horses in research institutions, both their social organization and their daily activity patterns must be taken into consideration. Free ranging (Berger, 1986) and wild (Boyd, 1998) horses live in bands composed of several mares, one or, occasionally, more stallions, foals and yearlings. The adult composition of these bands is quite stable. Members within established groups have preferred associates with whom they spend most of their time. This social organization should be respected as much as possible in equine housing, in order to safeguard both the welfare of the species and the validity of the data obtained. Horses communicate to one another and to us with their ears, tail and general posture, as well as vocally.

Feral (Salter and Hudson, 1979) and pastured domestic (Crowell-Davis et al., 1985) horses spend 40-60% of their time grazing both day and night (Keiper and Keenan, 1980). Grazing combines feeding with exercise because the pattern is for the horse to take a few steps, prehend, bite off and chew several tufts of grass and then take a few more steps before repeating the process. Horses rarely trot or canter unless startled, so exercise need not be strenuous to preserve their welfare. They stand up most of the time but must lie down for deep sleep (Ruckebusch, 1972); because this takes place in the early morning hours of darkness, one may not appreciate that the animals lie down and need a suitable surface to do so. Horses drink in association with feeding. Since they eat frequently they also drink frequently (Laut et al., 1985).

Horses may allogroom one another. The occurrence of this behavior varies among pairs of horses and with the season (Clutton-Brock et al., 1976). Other care of the coat includes biting at the coat, rolling, rubbing on inanimate objects or rubbing the face on the forelegs. Provision must be made for these activities to maintain the horse's welfare.

- **Pasture** is typically the best environment for horses because it allows them to graze (Figure 1) and live in a group (Figure 2). There must be a source of shelter from both the sun and from rain or snow. The fences must be adequate to restrain the animals but not dangerous. Barbed wire should not be used. Post and rail or board fences are traditional, but the latter require upkeep and may be eaten

Figure 1. Grazing [horse walks a few steps, prehends food with his head down, walks, etc.] is the major natural activity of horses.

Figure 2. Pasture with at least one other horse is the most species-appropriate environment.

Figure 3. Horses interacting across a gate. The nostril-to-nostril greeting is often followed by squealing and striking with the foreleg. The foreleg could become entrapped between the rails. This gate is safe because the rails are rounded.

by the horse. Woven wire topped by wood is suitable. Gates must be properly designed to avoid injury (Figure 3).

Unfortunately, there are not many experiments that can be performed with pastured horses, but pasture should be used for a holding area between experiments. Paddocks are suitable providing they are covered with quality grass. They are also valuable for exercise when the horses are released in a group. When allowed to control their own environment, horses will spend the majority of their time outside in a paddock rather than in a stall (Houpt and Houpt, 1988).

- In contrast to the natural social group, domestic horses, in general, and research horses, in particular, are kept either singly or in groups that frequently change composition. **Social isolation** is a disturbing experience for horses, and isolated subjects show behavioral and physiological stress reactions (Mal et al., 1991a,b) that may influence research data collected from them. When they have the option, horses will spend about half of their time in contact with other horses (Houpt and Houpt, 1988). Some experiments, particularly those dealing with nutrition, ectoparasites or infectious diseases, may require temporary isolation, an unnatural diet or no bedding. In such cases, environmental enrichment must be provided. The enrichment should be something, such as foraging, that simulates normal equine behavior. A food dispensing apparatus (Figures 4a & 4b) was first described by Malpass and Weigler (1994) and has been shown to be effective in maintaining normal behavior in stabled horses (Winskill et al., 1996).

  When horses are kept individually, they may be housed in box stalls (Figures 5 & 6) that allow them to turn around but usually prevent any social contact, or they may be housed in tie stalls (Figure 7) that are probably more species-appropriate because they allow social contact (McDonnell et al., 1999).

- Because horses who are confined for a long time develop edema, and because after confinement in a stall without the opportunity to **exercise**, horses show a compensatory increase in activity when released from their stalls (Houpt et al., 2001), horses should be released daily, preferably in a group on a pasture or paddock. When given a choice, horses spend twice as much time in a paddock when other horses are around as when they are released on it alone (Lee et al., 2001).

- Horses prefer **bedding** to a bare surface (Hunter and Houpt, 1989; Figure 7). The preference for bedding is most marked when the horses are lying down, which is not surprising because that is the time when their soft tissues are in contact with the substrate. Horses prefer straw to wood shavings in some studies (Mills et al., 2000), but not others (Hunter and Houpt, 1989). Wood shavings are more absorbent and more insulating than straw (Airaksinen et al., 2001) but don't provide a source of oral enrichment.

- Horses should have **water** available ad libitum. Automatic waterers are often provided, but, although they are labor saving, horses do not drink as much from automatic waterers as from buckets, probably because a deep receptacle is more natural for a suction drinker such as the horse (Nyman and Dahlborn, 2001). Automatic water lines may be broken by horses, and the waterers may malfunction, either depriving the horse of water or flooding the stall (see Figure 5, arrow). If automatic waterers are used, care must be taken that a newly introduced horse can operate them. It is always a good idea also to provide a bucket until the new animal has been observed to drink from the automatic waterer.

- There are several oral stereotypies in captive horses; wood chewing and cribbing are the most common. Diet can modify them. Wood chewing decreases when hay rather than pellets is fed (Willard et al., 1977). Cribbing occurs most frequently subsequent to eating grain—the sweeter the feed the more the horse cribs. Increasing the amount of hay or decreasing or eliminating sweet feed will reduce cribbing (Kusunose, 1992; Gillham et al., 1994). Providing ad libitum access to **hay** is the best way to avoid the development of cribbing and wood chewing in stalled horses (Figure 8).

- **Visual contact** with conspecifics can have positive or negative effects on the horse's comfort. Weaving—a locomotor stereotypic behavior—decreases in proportion to the numbers of sides of a stall that have windows (Cooper et al., 2000). Presumably, it is the sight of other horses that increases the horse's comfort because mirrors are also effective in decreasing weaving (McAfee et al., 2002). Negative effects are perception of threats from another horse and anxiety, when a horse who is usually present is removed. As a general rule, horses should have a view of other compatible horses.

- Horses will work for **light** in a dark stable (Houpt and Houpt, 1988). Their living quarters should, therefore, be well illuminated. Contrast in light between environments can be a problem in that horses often refuse to enter a barn that is markedly darker than the outdoor environment.

- **Grooming** lowers the heart rate of a horse, suggesting that it has a relaxing, calming effect (Feh and de Mazières, 1993). A stalled horse, however, is denied the opportunity to allogroom (Figure 9) and may, in addition, even be denied the opportunity to autogroom. Some horses are reluctant to roll in their stalls, and stalls often provide no

Figure 4a. A food dispensing cylinder [Pasture Pal®] that can be used as an enrichment device. The horse must push it to obtain pellets or grain from holes in the device (photo by Txema Peralta).

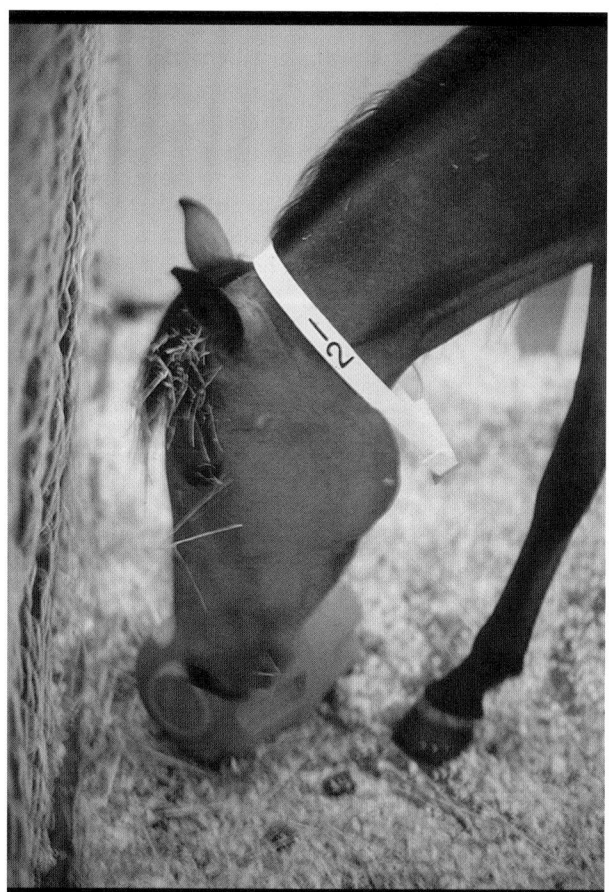

Figure 4b. Horse foraging from the Pasture Pal® (photo by Txema Peralta).

Figure 5. A box stall. This stall has the advantage that the horse cannot easily injure himself. The disadvantage is that the animal has social contact only distantly through two sets of bars with the horse across the aisle. Arrow indicates automatic waterer supply line.

Figure 6. This type of box stall allows visual and tactile contact between horses. Note that the dividing wall between the stalls has a screened window to allow more visual contact.

or ineffective rubbing surfaces. Grooming by humans is probably a suitable substitute.

Horses should be groomed several times a week. This is particularly important if they are dirty from rolling in the mud and when they are shedding in the spring. Grooming is also an opportunity to examine the horse's body; abnormalities may be detected that otherwise would not be noticed until they have become more serious. Hooves should be cleaned daily.

- Perhaps the most important contribution to a research horse's comfort is the **caretaker**. The caretaker should be willing and able to handle horses gently but effectively. He or she should be alert to the horse's environment [e.g., sharp edges where a horse can be injured or unlatched stalls], to changes in water and food intake and to social relations among the horses.

Since we usually do not keep horses in stable groups or pairs in research facilities, aggression frequently occurs even between horses in separate stalls or paddocks. If the aggressor is highly motivated or the pair evenly matched, even very sturdy fences can be kicked down, resulting in damage to the animals and/or the facility. For these reasons, caretakers have to be aware of the relationship between the horses in their charge and make sure that only compatible partners are placed next to each other. The caretaker can determine whether aggression is imminent if two horses, after touching nostrils (Figure 3), squeal and strike with a foreleg, if one or both partners approach with flattened ears or if their tails are lashing. It is important to watch the horses at feeding time because that is when aggression is most apt to occur. Another critical situation is when horses are being taken from a paddock or corral. Often, the animals will fight to be the first to leave. The best rule here is to release the most dominant horse [the one who can displace all the other horses with impunity] into the paddock last, and return him back into the stable first so that he won't prevent others from entering or leaving.

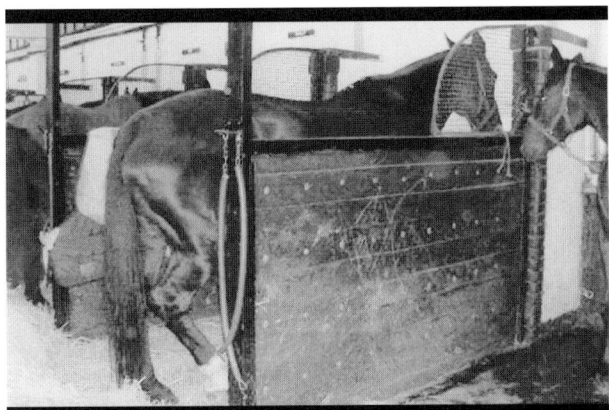

Figure 7. A Straight or tie stall with straw bedding. Note that there is a screen between the horses to allow social contact but not biting. The partitions should be solid, although ideally still removable in an emergency, as the horses will lean on them; wood is generally more suitable than metal even if it is harder to keep clean.

Figure 8. A mare and foal fed ad lib hay from a rack. Note that the ground around the rack is devoid of grass. Racks prevent wastage of hay but are not a natural way for a horse to eat (cf., Figure 1). Horses may inhale particulate matter including pathogens and allergens when the hay is above their chest.

Figure 9. Mutually grooming [allogrooming].

The **essentials of adequate housing** for horses in research institutions are: Access to hay and water ad libitum, a soft substrate for lying, visual contact with other horses and opportunity to exercise at will with other horses. Handling should involve firm and gentle restraint but not fear or pain. Grooming is necessary for comfort as well as cleanliness.

References

Airaksinen S, Heinonen-Tanski H, Heiskanen M-L 2001. Quality of different bedding materials and their influence on the compostability of horse manure. Journal of Equine Veterinary Science 2, 125-130

Berger J 1986. Wild Horses of the Great Basin The University of Chicago Press, Chicago, IL

Boyd L 1998. The 24-h time budget of a Takhi harem stallion (*Equus ferus przewalskii*) pre- and post-reintroduction. Applied Animal Behaviour Science 60, 291-299

Clutton-Brock TH, Greenwood PJ, Powell RP 1976. Ranks and relationships in Highland ponies and Highland cows. Zeitschrift für Tierpsychologie 41, 202-216

Cooper JJ, McDonald L, Mills DS 2000. The effect of increasing visual horizons on stereotypic weaving: Implications for the social housing of stabled horses. Applied Animal Behaviour Science 69, 67-83

Crowell-Davis SL, Houpt KA, Carnevale J 1985. Feeding and drinking behavior of mares and foals with free access to pasture and water. Journal of Animal Science 60, 883-889

Feh C, de Mazières J 1993. Grooming at a preferred site reduces heart rate in horses. Animal Behaviour 46, 1191-1194

Gillham SB, Dodman NH, Shuster L, Kream R, Rand W 1994. The effect of diet on cribbing behaviour and plasma endorphin in horses. Applied Animal Behaviour Science 41, 147-153

Houpt KA, Houpt TR 1988. Social and illumination preferences of mares. Journal of Animal Science 66, 2159-2164

Houpt KA, Houpt TR, Johnson JL, Erb HN, Yeon SC 2001. The effect of exercise deprivation on the behaviour and physiology of straight stall confined pregnant mares. Animal Welfare 10, 257-267

Hunter L, Houpt KA 1989. Bedding material preferences of ponies. Journal of Animal Science 67, 1986-1991

Keiper RR, Keenan MA 1980. Nocturnal activity patterns of feral horses. Journal of Mammalogy 61, 116-118

Kusunose R 1992. Diurnal pattern of cribbing in stabled horses. Japanese Journal of Equine Science 3, 173-176

Laut JE, Houpt KA, Hintz HF, Houpt TR 1985. The effects of caloric dilution on meal patterns and food intake of ponies. Physiology and Behavior 35, 549-554

Lee J, Floyd T, Houpt K 2001. Operant and two-choice preference applied to equine welfare. Proceedings of the 35th International Society of Applied Ethology International Congress, 110

Mal ME, Friend TH, Lay DC, Vogelsang SG, Jenkins OC 1991a. Behavioral responses of mares to short-term confinement and social isolation. Applied Animal Behaviour Science 31, 13-24

Mal ME, Friend TH, Lay DC, Vogelsang SG, Jenkins OC 1991b. Physiological responses of mares to short term confinement and social isolation. Equine Veterinary Science 11, 96-102

Malpass JP, Weigler BJ 1994. A simple and effective environmental enrichment device for ponies in long-term indoor confinement. Contemporary Topics in Laboratory Animal Science 33, 74-76.

McAfee LM, Mills DS, Cooper JJ 2002. The use of mirrors for the control of stereotypic weaving behaviour in the stabled horse. Applied Animal Behaviour Science (in press)

McDonnell SM, Freeman DA, Cymbaluk NF, Schott HC, Hinchcliff KW, Kyle B 1999. Behavior of stabled horses provided continuous or intermittent access to drinking water. American Journal of Veterinary Research 60, 1451-1456

Mills DS, Eckley S, Cooper JJ 2000. Thoroughbred bedding preferences, associated behaviour differences and their implications for equine welfare. Journal of Animal Science 70, 95-106

Nyman S, Dahlborn K 2001. Effect of water supply method and flow rate on drinking behavior and fluid balance in horses. Physiology and Behavior 73, 1-8

Ruckebusch Y 1972. The relevance of drowsiness in the circadian cycle of farm animals. Animal Behaviour 20, 637-643

Salter RE, Hudson RJ 1979. Feeding ecology of feral horses in western Alberta. Journal of Range Management 32, 221-225

Willard JG, Willard JC, Wolfram SA, Baker JP 1977. Effect of diet on cecal pH and feeding behavior of horses. Journal of Animal Science 45, 87-93

Winskill LC, Waran NK, Young RJ 1996. The effect of a foraging device (a modified 'Edinburgh Foodball') on the behaviour of the stabled horse. Applied Animal Behaviour Science 48, 25-35

Dr. K.A. Houpt graduated from the University of Pennsylvania with a VMD and PhD. She is a diplomate of The American College of Veterinary Behaviorists and director of the Animal Behavior Clinic at Cornell University College of Veterinary Medicine. Her main research focus is equine welfare.

Dr. T.S. Ogilvie-Graham's credentials include DVM&S in Applied Equine Ethology from Edinburgh University. He is Lieutenant Colonel RAVC, in charge of the Royal Household Cavalry.

Comfortable Quarters for Chickens in Research Institutions

Detlef W. Fölsch, Marlene Höfner, Marion Staack and Gerriet Trei
University of Kassel, Faculty of Agriculture, International Rural Development and Environmental Protection, Department of Animal Behavior and Management, Nordbahnhofstraße 1a, D-37213 Witzenhausen, Germany

The progenitors of the domestic chickens—the red jungle fowl of southeast Asia—are social birds who prefer habitats with dense vegetation that provides good cover. Neither thousands of years of domestication nor the recent extreme selective breeding for productivity have significantly altered the biological and behavioral characteristics of these birds (Fölsch and Vestergaard, 1981; Rogers, 1995). This has to be kept in mind when suitable housing for chickens is designed.

The different behaviors shown by chickens can be categorized as follows:

- **Foraging behavior** consists of pecking and ground scratching followed by ingestion. The animals use their beaks, which are richly innervated, to search for and to retrieve food. Chickens spend 35 to 50% of the day scratching and pecking for food. Under natural conditions they consume many different food items such as seeds, fruits, grass, insects, worms and berries, but also little stones [calcium carbonate and calcium silicate], which are important for the digestive process in the chicken's stomach as well as for the formation of the skeleton and the eggshell. If they do not spend a major portion of the day engaged in foraging activities, chickens tend to peck, pull and tear at objects or conspecifics, and often develop feather pecking behavior. Feather pecking is a redirected ground pecking (Blokhuis and Arkes, 1984). It is a behavioral disorder, a sign that the housing and feeding conditions are not corresponding to the animals' behavioral needs (Huber-Eicher and Wechsler, 1997). It goes without saying that partial beak amputation is not an ethically acceptable remedy to feather pecking because this procedure results in considerable pain that persists for several weeks (Gentle et al., 1990).
- **Locomotive behavior** includes walking, running, flying and wing flapping. Hens will walk about 1 to 1.5 km per day and fly to and from elevated places if they have the opportunity to do so (Keppler and Fölsch, 2000).
- **Resting behavior** includes standing, lying, sleeping and dozing. Chickens prefer to roost on higher rather than lower perches (Blokhuis, 1984). They probably do this for safety reasons, to be out of reach of ground predators. Access to perches during rearing attenuates the flightiness of mature animals (Brake, 1987).
- **Maintenance-Comfort behavior** consists of preening, stretching, flapping, dustbathing, sunbathing and body shaking. To keep their feathers in good condition, chickens

Figure 1. Chickens spend 35 to 50% of the day scratching and pecking for food.

Figure 2. Comfortable laying nests are a "must" for any chicken quarters (photo by Viktor Reinhardt).

Comfortable Quarters for Laboratory Animals Reinhardt V, Reinhardt A (eds), 101-108. Animal Welfare Institute, Washington, DC 20007

Figure 3. For resting, sleeping and withdrawal, chickens prefer elevated places, natural trees and bushes. This is why the provision of perches is an imperative for any housing system of chickens.

must be able to take dustbaths regularly and preen themselves. When dustbathing, chickens toss the litter onto and between the fluffed feathers and subsequently enclose it by flattening the feathers. This comfort behavior regulates the amount of feather lipids and maintains down and feather structure in good conditions. Poor plumage condition negatively influences animal welfare (van Liere and Bokma, 1987), and hence the validity of research data. Chickens will frequently sunbathe if they are given the opportunity. Daylight controls and triggers many of their physiological processes. It also stimulates their metabolism, plays an important part in the formation of red and white blood cells and of vitamin D, and promotes the secretion of hormones necessary for growth and reproduction.

- **Social behavior** includes pecking, threatening, chasing, kicking, fighting, avoiding, crouching and vocalizing. In the wild, hens and cocks of different ages live in small groups and form a cohesive community. One cock lives in a group with about seven hens (Figure 1). The social structure of a flock depends on the physiological, psychological and physical state of each member and is influenced by the appearance of the individual [e.g., animal is ill, injured, moulting, brooding]. A stable rank order is formed within a small group of chickens on the basis of personal affiliations, threat and avoidance behavior, and factors such as age, color, sex and the size of the comb (Keppler et al., 1997). Social interactions can be friendly, for example a cock calling his hens to a food source, or they can be agonistic, for example one hen chasing another hen away from a limited food source. When it is possible, chickens will seek shelter for protection from predators and aggressive conspecifics.

- All hens show elements of the typical **nesting-and-laying behavior sequence**: separating from the flock, examining potential nest sites, scratching and pecking at nest material, building a nest or choosing an already formed nest, entering the nest, forming a hollow, laying an egg, rolling the egg under the body, lying on the egg, getting up, standing, leaving the nest and cackling. If no adequate nest site is available, hens will develop abnormal nesting and laying behavior and possibly stereotyped activity patterns. Hens prefer to lay their eggs in sheltered places where manipulable material is available (Figure 2). They have an urgent demand for suitable nesting material, such as straw, and will readily "work" to gain access to it (Huber et al., 1985; Gunnarsson et al., 2000).

Figure 4a, b. Chicks and young hens should get perches early in life so that they learn perching and using the third dimension.

Figure 5a, b. A dustbath is used for care and cleaning of plumage and enhances well-being. Chickens prefer to dustbathe in groups. If the dustbath is provided outside, it should be roofed and should give protection from drafts. Quartz sand with charcoal and flowers of sulphur is recommended as a dustbathing substrate.

- Chickens are easily frightened by sudden changes in their environment and by exposure to human beings who, after all, are natural predators. The birds show typical **antipredator responses** such as alarm-calling, freezing, panic attempts to run away, vigorous struggling and feigning death. The provision of a more complex, enriched environment decreases fearfulness (Jones, 1982; Grigor et al., 1995).

Whether chickens are kept in rooms or in so-called enriched cages, the following provisions must be made to address the animals' basic species-specific needs: social housing, laying nest, elevated perch(es), natural light, area(s) for pecking, scratching and sand bathing (Figures 2-5).

A chicken should never be kept alone. Separation from conspecifics is a distressing experience (Rajecki et al., 1977) that is bound to have an influence on research data collected. The occurrence of feather-, foot- and claw damage is less in cages equipped with a nest box, a perch and an area for pecking/scratching/bathing, indicating that these items significantly improve the welfare of chickens (Appleby et al., 1993). As far as the hens' immediate response to their environment is concerned, space *per se* is not the only relevant factor. Just as, if not more, important are the structuring of the vertical dimension and the quality of the flooring. Hens will voluntarily "work" for access to litter, indicating that a litter substrate is highly valued by them (Matthews et al., 1994; Widowski and Duncan, 2000). Chickens should always have access to a sufficiently spacious litter area where they can move around without experiencing the discomfort associated with wire floors. Not surprisingly, chickens show a clear preference for litter over wire floors (Hughes, 1976). Hens with access to litter in an otherwise barren environment spent around 18% of their time in litter-related activities (Hughes and Channing, 1998).

A room may be readily transformed to suitable housing for chickens by placing a wire-mesh covered dropping pit on one side of the room and installing perches along the wall at different heights over the pit. The horizontal distance between two perches should be at least 35 cm (Figure 3). Chickens prefer roosts that are large [5 cm diameter] rather than small, and square or round rather than triangular in shape (Muiruri et al., 1990). The total perch-length is determined by the number of chickens and should be no less than 18 cm per animal. Chickens kept in unstructured enclosures will use locations near walls more than expected and avoid the empty central area (Newberry and Hall, 1990). The provision of different horizontal levels (Fölsch et al., 1988) or vertical panels can ameliorate this problem. Such structures facilitate a more uniform distribution of the birds, thereby increasing their useable floor space (Cornetto and Estevez, 2001) and, probably, also enhancing their well-being (Newberry and Shackleton, 1997).

A scratching area is an imperative to allow chickens to exhibit species-typical foraging behavior. Similar to other laboratory animals, the foraging drive is so strong in chickens that they will "work" for food in the presence of freely accessible identical food (Duncan and Hughes, 1972). This suggests that foraging is a rewarding activity in itself independent of caloric intake (Neuringer, 1969). A thin layer of sand covered by approximately 10 cm of chopped straw provides a suitable substratum onto which the grain can be scattered, thus triggering prolonged foraging. This scratching/foraging area should take up at least half of the horizontal floor space of the enclosure.

Hens need adequate nest boxes, preferably with manipulable material, like oat husks or chopped straw (Figure 2). One nest box per 5 hens should be

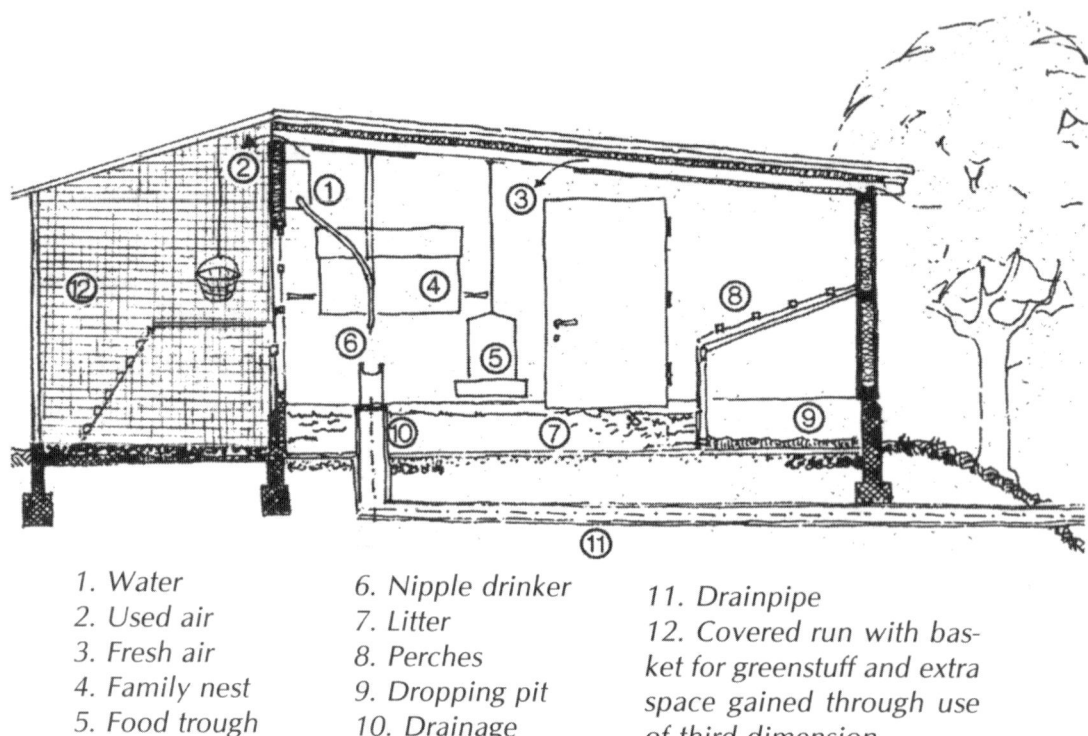

1. Water
2. Used air
3. Fresh air
4. Family nest
5. Food trough
6. Nipple drinker
7. Litter
8. Perches
9. Dropping pit
10. Drainage
11. Drainpipe
12. Covered run with basket for greenstuff and extra space gained through use of third dimension

Figure 6. Suggestions for hen house with covered bad-weather run.

provided. Its dimensions should be approximately 40 x 40 cm. If larger family nests are used, a nesting area of 1 m^2 per 50 hens is recommended.

It is said that group size should not exceed 80 animals, as chickens are only able to distinguish between 40 to 80 members of their own species. Nevertheless, there is evidence that hens kept in aviary units of 2,200 establish a social organization (Keppler et al., 1997). Stocking density should not exceed 5 birds per square meter of available surface area to avoid stress from overcrowding.

As with other laboratory animals, the presence of a social companion has a stress buffering effect in chickens (Jones and Merry, 1988). In order to avoid data-biasing stress responses, it is therefore advisable to allow individual chicks or hens to maintain visual and acoustic contact with at least one familiar conspecific during experimental procedures.

A variety of food should be offered to the chickens. If only meal or pellets are fed, the animals consume their ration too fast and do not spend enough time foraging. Feather pecking can easily develop under such conditions. To prevent this, grain should be thrown onto the litter to promote scratching and pecking behaviors. Fruit, grass, straw and hay should be provided in racks or in baskets hanging from the ceiling, so that the animals can pull and peck at the contents rather than at each other. Rearing chicks with access to sand, peat or straw as litter substrates for dustbathing, pecking and foraging reduces later tendencies to engage in feather pecking (Nørgaard-Nielsen et al., 1993; Huber-Eicher and Wechsler, 1997; Nicol et al.,

2001). It is, therefore, advisable to offer chicks access to litter from day one on (Huber-Eicher and Sebö, 2001). If straw is provided as litter substrate, special attention should be paid to its form, as long-cut straw is more efficient in reducing feather pecking than straw in shredded form (Wechsler and Huber-Eicher, 1998).

Feeding space should be 5 cm per bird on circular feeders and 12 cm per bird on feeding belts. There should be one watercup or one nipple drinker per 8 birds and a watertrough length of 2-3 cm per bird, respectively. Chickens can drink much better from open receptacles than from nipple drinkers. It is therefore recommended to provide water in troughs or cups rather than forcing the birds to drink from nipple drinkers.

Chickens should have access to a box filled with sand so that they can take dustbaths (Figure 5). Dustbathing is a social activity that is usually performed by several birds at the same time. The sand box has to be relatively spacious, i.e., 80 x 80 cm per 50 birds.

A bad-weather run should be provided so that the chickens have exposure to natural daylight and seasonal temperature variations throughout the year. The run should have a roof so that the birds have access to the sheltered area even in bad weather. Wire-mesh walls or a meshed plastic netting will protect the animals from predators. The sheltered run should be about half the size of the hen house and have a concrete floor covered with a layer of straw and sand (Figure 6-7).

Chickens clearly prefer an outside run to a cage, once initial unfamiliarity with the run has been overcome

Figure 7. A bad-weather run should be provided so that the chickens have exposure to natural daylight and seasonal temperature variations throughout the year.

(Dawkins, 1977). An outdoor run is, therefore, highly recommended. It should be covered to protect the animals from prey birds or provisioned with bushes and trees so that the animals have places where they can take cover (Figure 8). The addition of organic material or a compost heap gives the chickens optimal opportunities to perform a wide array of species-specific behaviors. The stocking density in the outdoor run may be up to 10 hens/m^2 if pastures are used alternately. It should be no more than 2 hens/m^2 if no rotation of pastures is possible. Animals who spend more than half of their time outside are less fearful, and the time of tonic immobility is shorter compared with animals who have no access to an outdoor run (Grigor et al., 1995). Also, hens who have been kept in free range systems show less fear reactions after being transported than hens who have been kept in cages (Scott et al., 1998). These findings indicate keeping chickens in free range systems fosters their capability to adapt to potentially stressful situations.

A cage is not an appropriate housing environment for chickens. The birds demonstrate this by exhibiting more pronounced fear responses when kept in cages versus floor pens (Jones and Faure, 1981a). It should be emphasized again that any cage for chickens has to be equipped at a minimum with a laying nest, a high perch, and an area for pecking, scratching and dustbathing to meet fundamental behavioral needs of the animals (Figure 9).

If chickens are kept in cages in a research institution [e.g., for metabolic studies], the cages should not be arranged in multiple tiers in order to avoid data-confounding variables resulting from different degrees of

Figure 8. An outside run is highly recommended. It should have bushes and trees for cover.

fear responses in birds kept at different levels of the room (Sefton, 1976; Jones, 1985; Hemsworth et al., 1993) and different illumination in cages at different distances from the light source.

The greatest risk of physical injury will occur if chickens become frightened and attempt to escape from their cages, either during catching procedures or simply when disturbed by human presence. It is, therefore, important to allow sufficient space for running and wing flapping to maintain good bone strength (Whitehead et al., 1997/98). This freedom must be coupled with the provision of a small, safe catching

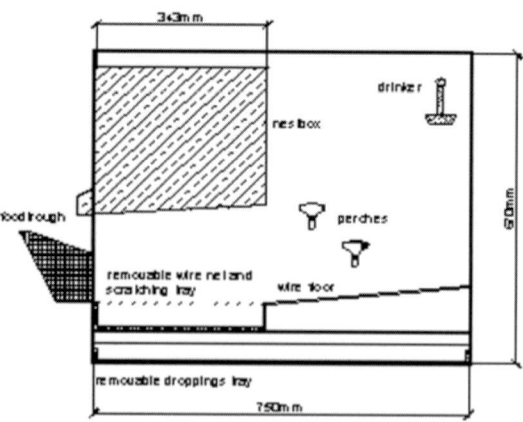

Figure 9. Cage design addressing the minimum behavioral needs of the chickens kept in research institutions. Note that provision is made for a nestbox, a scratching area, perches and a water receptacle.

area. The birds can often be enticed into such areas if these are well lit while the rest of the room is temporarily darkened (Nicol, 1995).

Chickens are highly susceptible to stress when they are caught and handled. Regular gentle interactions with chickens reduces fear and facilitates easier, i.e., less stressful handling during procedures (Hughes and Black, 1976; Jones and Faure, 1981b; Gross and Siegel, 1982; Jones, 1994; Figure 10). The stress response can further be minimized by performing any procedures during dim-lighting conditions. At dusk, dawn, or when lights are turned off, chickens can normally be picked up from the floor or from perches without causing undue commotion. A chicken should always be carried by holding both legs while supporting the body with the hands. Gently tucking the head under the arm helps to keep the animal calm.

References

Appleby MC, Smith SF, Hughes BO 1993. Nesting, dust bathing and perching by laying hens in cages: Effects of design on behaviour and welfare. British Poultry Science 34, 835-847

Blokhuis HJ 1984. Rest in poultry. Applied Animal Behaviour Science 12, 289-303

Blokhuis HJ, Arkes JG 1984. Some observations on the development of feather pecking in poultry. Applied Animal Behaviour Science 12, 145-157

Brake J 1987. Influence of presence of perches during rearing on incidence of floor laying in broiler breeders. Poultry Science 66, 1587-1589

Figure 10. Tame chickens are more reliable research models because they show little or no stress responses during handling procedures (photo by Annie Reinhardt).

Cornetto TL, Estevez I 2001. Influence of vertical panels on use of space by domestic fowl. Applied Animal Behaviour Science 71, 141-153

Dawkins M 1977. Do hens suffer in battery cages? Environmental preferences and welfare. Animal Behaviour 25, 1034-1046

Duncan IJH, Hughes BO 1972. Free and operant feeding in domestic fowls. Animal Behaviour 20, 775-777

Fölsch DW, Vestergaard K 1981. The Behaviour of Fowl—The Normal Behaviour and the Effect of Different Housing Systems and Rearing Methods—Animal Management Volume 12 Birkäuser, Basel, Switzerland

Fölsch DW, Huber H-U, Bölter U, Gozzoli I 1988. Research on alternatives to the battery systems for laying hens. Applied Animal Behaviour Science 20, 29-45

Gentle MJ, Waddington D, Hunter LN, Jones B 1990. Behavioural evidence for persistent pain following partial beak amputation in chicken. Applied Animal Behaviour Science 27, 149-157

Grigor PN, Hughes BO, Appleby MC 1995. Effects of regular handling and exposure to an outside area on subsequent fearfulness and dispersal in domestic hens. Applied Animal Behaviour Science 44, 47-55

Gross WB, Siegel PB 1982. Socialization as a factor in resistance to infection, feed efficiency, and response to antigen in chickens. American Journal of Veterinary Research 43, 2010-2012

Gunnarsson S, Matthews LR, Foster TM, Temple W 2000. The demand for straw and feathers as litter substrates by laying hens. Applied Animal Behaviour Science 65, 321-330

Hemsworth PH, Barnett JL, Jones RB 1993. Situational factors that influence the level of fear of human by laying hens. Applied Animal Behaviour Science 36, 197-210

Huber H-U, Fölsch DW, Stähli U 1985. Influence of various nesting materials on nest site selection of the domestic hen. British Poultry Science 26, 367-373

Huber-Eicher B, Wechsler B 1997. Feather pecking domestic chicks: Its relation to dustbathing and foraging. Animal Behaviour 54, 757-768

Huber-Eicher B, Sebö F 2001. Reducing feather pecking when raising laying hen chicks in aviary systems. Applied Animal Behaviour Science 73, 59-68

Hughes BO 1976. Preference decisions of domestic hens for wire or litter floors. Applied Animal Ethology [Applied Animal Behaviour Science] 2, 155-165

Hughes BO, Black AJ 1976. The influence of handling on egg production, egg shell quality and avoidance behaviour of hens. British Poultry Science 17, 135-144

Hughes BO, Channing CE 1998. Effects of restricting access to litter trays on their use by caged laying hens. Applied Animal Behaviour Science 56, 37-45

Jones RB 1982. Effects of early environmental enrichment upon open-field behavior and timidity in the domestic chick. Developmental Psychobiology 15, 105-111

Jones RB 1985. Fear responses of individually-caged laying hens as a function of cage level and aisle. Applied Animal Behaviour Science 14, 63-75

Jones RB 1994. Regular handling and the domestic chick's fear of human beings: generalisation of response. Applied Animal Behaviour Science 42, 129-143

Jones RB, Faure JM 1981a. Tonic immobility («righting time») in laying hens housed in cages and pens. Applied Animal Ethology 7, 369-37

Jones RB, Faure JM 1981b. The effects of regular handling on fear responses in the domestic chick. Behavioural Processes 6, 135-143

Jones RB, Merry BJ 1988. Individual or paired exposure of domestic chicks to an open field: some behavioural and adrenocortical consequences. Behavioural Processes 16, 75-86

Keppler C, Schnurrenberger-Bölter U, Fölsch DW 1997. Activity and social relationships of chickens (Gallus gallus f. domesticus) in aviary systems—methods and preliminary results. In 5th Symposium on Poultry Welfare, Koene P, Blokhuis HJ (eds), 105-106. World's Poultry Science Association, University of Wageningen, Netherlands

Keppler C, Fölsch DW 2000. Locomotive behaviour of hens and cocks (Gallus gallus f. domesticus): Implications for housing systems. Archiv für Tierzucht 43, 184-188

Matthews LR, Temple W, Foster TM, McAdie TM 1994. Quantifying the environmental requirements of layer hens by behavioural demand functions. Applied Animal Behaviour Science 40, 91

Muiruri HK, Harrison PC, Gonyou HW 1990. Preferences of hens for shape and size of roosts. Applied Animal Behaviour Science 27, 141-147

Neuringer AJ 1969. Animals respond for food in the presence of free food. Science 166, 399-401

Newberry RC, Hall JW 1990. Use of pen space by broiler chickens: effects of age and pen size. Applied Animal Behaviour Science 25, 125-136

Newberry RC, Shackleton DM 1997. Use of visual cover by domestic fowl: a Venetian blind effect? Animal Behaviour 54, 387-395

Nicol CJ 1995. Environmental enrichment for birds. AWIC Resource Series No. 2—Environmental Enrichment Information Resources for Laboratory Animals 1995-1995, Birds, Cats, Dogs, Farm Animals, Ferrets, Rabbits, and Rodents, 1-3

Nicol CJ, Lindberg AC, Phillips AJ, Pope SJ, Wilkins LJ, Green LE 2001. Influence of prior exposure to wood shavings on feather pecking, dustbathing and foraging in adult laying hens. Applied Animal Behaviour Science 73, 141-155

Nørgaard-Nielsen G, Vestergaard K, Simonsen HB 1993. Effects of rearing experience and stimulus enrichment on feather damage in laying hens. Applied Animal Behaviour Science 38, 345-352

Rajecki DW, Suomi SJ, Scott EA, Campbell B 1977. Effects of social isolation and social separation in domestic chicks. Developmental Psychology 13, 143-155

Rogers LJ 1995. The Development of Brain and Behaviour in the Chicken CAB International, Wallingford, UK

Scott GB, Connell BJ, Lambe NR 1998. The fear levels after transport of the hens from cages and a free-range system. Poultry Science 77, 62-66

Sefton AE 1976. The interactions of cage size, cage level, social density, fearfulness and production of Single Comb White Leghorns. Poultry Science 55, 1922-1926

van Liere DW, Bokma S 1987. Short-term feather maintenance as a function of dust-bathing in laying hens. Applied Animal Behaviour Science 18, 197-204

Wechsler B, Huber-Eicher B 1998. The effect of foraging material and perch height on feather pecking and feather damage in laying hens. Applied Animal Behaviour Science 58, 131-141

Whitehead C, Fleming B, Bishop S 1997/98. Towards a Genetic Solution to Osteoporosis in Laying Hens: Annual Report Roslin Institute, Midlothian, UK

Widowski TM, Duncan IJH 2000. Working for a dustbath: are hens increasing pleasure rather than reducing suffering? Applied Animal Behaviour Science 68, 39-53

Detlef W. Fölsch is Professor of Farm Animal Ethology and Management at the University of Kassel/Witzenhausen, Germany. His extensive research in chicken ethology and commitment to animal welfare and responsible agriculture were instrumental in the implementation of the ban on battery cages for laying hens in Switzerland in 1986. He and his co-workers were also instrumental in adoption of a German law mandating that, beginning in 2007, laying hens will have to be kept in systems other than battery cages.

Marlene Höfner is a graduate agriculturist. She investigated the effects on optimizing outside runs on the behavior of hens and on the environment at the University of Kassel/Witzenhausen, Germany.

Marion Staack is a graduate agriculturist with a Master of Science degree in Applied Animal Behavior and Animal Welfare. She has collaborated with Detlef Fölsch in the project "Alternative Housing Systems for Poultry."

Gerriet Trei is a Ph.D. student with Professor Fölsch. He works on feeding behavior, optimizing the composition of food for hens. He has collaborated with Detlef Fölsch in the projects "Alternative Housing Systems for Poultry in Hessia" and "Animal and Environmental Friendly Housing Systems for Poultry" of the German Ministry for Consumer Protection, Food and Agriculture.

Comfortable Quarters for Amphibians and Reptiles in Research Institutions

Michael D. Kreger

Division of Scientific Authority, U.S. Fish and Wildlife Service, 4401 North Fairfax Drive, Arlington, VA 22203, USA

There are approximately 6000 species of reptiles and 4000 species of amphibians. Some are completely aquatic, some rarely leave the trees, and some are burrowers. They are found in almost every habitat on the planet. At first glance, it would appear difficult to list criteria for laboratory housing for such a diverse group of animals. However, there are three general keys to successful housing:

1. a knowledge of the biology of the specific species including the basic, but essential needs of ectotherms;
2. an ability to replicate the most important features of the reptile's or amphibian's natural environment in the housing and care provided to the animals in the laboratory; and
3. caretakers who are able to recognize signs of discomfort, stress, and ill health in the particular species.

- Reptiles and amphibians are **ectotherms** [cold-blooded animals]. Unlike endotherms [warm-blooded animals], their body temperature is strictly dependent on the ambient environment.

  The advantage of ectothermy is that the resting metabolic rate and general energy requirements are less than those for mammals or birds of comparable size since no metabolic energy is spent on warming or cooling the body, and less energy is spent on searching for prey because less food is required to meet the body's low energy demands. The disadvantage of ectothermy, however, is that the ambient temperature determines the animal's metabolic processes and behavior. The animal must actively seek temperatures that will allow him or her to feed, digest food, hibernate, etc. Reptiles and amphibians literally "select" their body temperature by finding the appropriate thermal environment through basking, burrowing, hiding under logs or leaves, or entering water. For example, after a meal, snakes will move towards a heat source to aid digestion, and they will retreat to a cooler area following defecation.

  In many respects cold-blooded animals are more interactive with their environments than warm-blooded animals. At the same time, they tend to have greater problems adapting to changes in their species-typical

  environment (Warwick, 1987; cf. Wright, 1994). Therefore, the design of their artificial habitats demands special care if research-biasing stress and distress responses to species-inadequate environmental conditions are to be avoided. "Whether an observer maintains a high personal respect of the well-being of the individual animal or holds classic concepts of animals as being experimental 'models,' it should be more widely recognized that there is typically a scientific necessity to have animals at ease with their environments if studies are to remain objective" (Warwick, 1990a, p. 363).

- The knowledge of the thermal limits of a species is a basic condition for its proper care. Individual animals must be observed regularly and carefully to assure that their microhabitat suits their thermal requirements. If a reptile or amphibian spends all the time under or on the heat source, the ambient **temperature** is—obviously—too cool. If the animal stays as far away as possible from the heat source, the temperature is too warm.

  Temperature is best timer-controlled, taking natural temperature gradients [evening temperatures drop significantly in the desert for example] at the individual vivarium or tank level into account. Depending on the size of the enclosure, a gradient can be established by using either radiant heat from quartz heaters used to brood chicks, 25 to 250 watt incandescent light bulbs placed

Comfortable Quarters for Laboratory Animals Reinhardt V, Reinhardt A (eds), 109-114. Animal Welfare Institute, Washington, DC 20007

Figure 1.* Broad-spectrum fluorescent and black light tubes are placed over the screened top of this reptile enclosure. The heat lamp is provided on the side near the perches where the lizard basks. The plastic on the cage sides helps increase humidity.

- The quality of **light** is important for the general health of diurnal [active during the day] species. Sunlight, the best form of illumination, is usually not available in the research laboratory setting and, if available, the glass from windows and tanks blocks its beneficial ultraviolet [UV] rays. Animals who do not eat whole prey [such as herbivorous lizards] particularly rely on UV light to induce vitamin D_3 production, which is necessary for calcium metabolism. UV light must therefore be provided artificially. In general, a long wavelength UV lamp [black light bulb] in combination with two or three broad-spectrum fluorescent tubes [e.g., Vita-Lite®] is recommended for reptiles and larval amphibians such as tadpoles (Figure 1). To avoid burns, the animals should not have direct access to the bulbs which, however, must be placed no more than four feet above them. No special lighting is required for nocturnal species although red, blue, or black lights can be used to illuminate the animals for the researcher.

- Different species have different needs in terms of **relative humidity** and water availability. Humidity can be controlled at the room level if all animals are from the same habitat. Humidifiers can be used for this purpose. In individual vivaria, humidity can be raised by evaporating water from a container placed near the heat source. Relative humidities should be maintained above 70%—preferably at 80%—for nearly all species of amphibians and reptiles (Pough, 1991).

  Room air changes should be limited to maintain high air humidity. Because amphibians and reptiles are ectotherms, their metabolic rates are low, as are their rates of production of carbon dioxide, urine, and feces. Consequently, high rates of air flow are not needed to renew the oxygen concentration of air in the animal room or to dissipate odors. Flow rates of one or two air changes per hour are desirable (Pough, 1992).

- Amphibians and reptiles benefit from daily water spraying with a plant mister. Water is absolutely critical to amphibians. They reproduce, their eggs hatch, and their larvae grow in **water** or moist environments. Although they do not drink, amphibians absorb water through the skin. They also breathe through membranes on the skin surface that must be kept moist. Since the skin is highly permeable to moisture and gases, amphibians become overheated and desiccated easily if not provided a moist environment. As a general rule, all amphibians need to have access to a clean water source that is large enough to allow them to submerge. Their permeable skin puts amphibians at risk of absorption of noxious substances. All their water must, therefore, be filtered, dechlorinated, and oxygenated or changed daily to reduce the possibility of putrefaction.

  For reptiles, water helps loosen skin about to be shed, it also helps body temperature regulation, and it is a drinking source. Snakes should be provided with a fresh standing water container large enough to submerge. Most terrestrial reptiles will drink from a small water bowl/dish. Some lizards, however, will not drink from a dish because they are biologically adapted to lapping water from the surface of leaves. For them as well as for very small snakes, a petri dish holding a water soaked sponge, or filled with a layer of absorbent cotton soaked with water

outside (Figure 1), or substrate heat such as heating pads, coils, or tapes placed under the enclosure. Relatively cool retreats such as shelter boxes [for reptiles] and/or leaf litter [for toads, frogs, and salamanders] should be provided to forestall overheating. Ventilation ports must be properly screened to prevent escape. There should be a mechanism to partially or completely block the ventilation ports to assist humidity regulation. Optimal temperatures for most reptiles are in excess of 77°F [25°C] and many, especially lizards, have optimal temperatures between 95 and 104°F [35-40°C] (Gilman, 1984a).

- In the laboratory, **day length** can be set for the entire room. It should approximate that found in the animals' natural habitat. Artificially controlled day length, in combination with artificially controlled microhabitat temperature, should simulate natural diurnal cycles and annual seasons, both of which affect the individual animal's feeding behavior, reproductive cycle, and torpor. Jones (1978) gives a useful formula and presents tables for the average day length for any latitude on any week. If an investigator is working with nocturnal species, the light cycle can be reversed so that the animals are active during the investigator's day.

Figure 2.* A young Komedo dragon on a tree stump. Many reptiles and amphibians use both vertical and horizontal space in their habitat. A small tank for tree-dwelling species with a complex, enriched vertical dimension that includes perches, foliage, etc. has more utilizable space than a larger tank with barren horizontal space.

Figure 3.* Rhinoceros iguana basking. Note rocks and shelter at the back. Design factors such as a rock for rubbing during ecdysis (shedding), and retreats will give the animals some control over their environments.

Many reptiles and amphibians use both vertical and horizontal space in their habitat. A small tank for tree-dwelling species with a complex, enriched vertical dimension that includes perches, foliage, etc. (Figure 2) has more utilizable space than a larger tank with barren horizontal space.

Furnishings can be simple, washable plastic such as PVC tubes. Materials that may leach chemical contaminants [e.g., dyes, fire retardants] must be avoided. Other design factors such as grooved "haul-out" sites for turtles, rough objects—such as a brick or a rock—for rubbing during ecdysis [shedding], basking/perching sites, and retreats (Figure 3) will give the animals some control over their environments by choosing locations according to their preferences (Figure 4). All cage props should be exchanged regularly or cleaned with a dilute bleach and water solution followed by rinsing with water and air drying until the residual chlorine has evaporated.

Shredded newspaper makes ideal burrowing material for some species of reptiles. Leaf litter and potting soil are suitable substrates but must be changed frequently. Sheets of newspaper make a highly recommended floor covering because they are clean, absorbent, inexpensive, and easy to change. Astroturf is more aesthetic; it can be easily sanitized and also provides a good rubbing material for shedding reptiles. To increase humidity, moss is a recommended substrate for small amphibians because it holds dampness well and provides cover. Moss and other organic material must be changed frequently to minimize bacterial and fungal growth. Other floorings like

Figure 4.* Lizard basking on a branch. Note that water, multiple basking sites and cover are provided.

will provide an adequate water source (Gilman, 1984a). All water must be clean and dechlorinated, and the water receptacles designed in such a way that they cannot be turned over by the animal(s).

- Most of the basic husbandry needs of reptiles and amphibians can be met with what, to human eyes, may seem like a stark and sterile **enclosure**. Plastic bins, aquarium tanks, polyurethaned wood and Plexiglas enclosures work very well.

  The size of the enclosure depends on the size and activity of the individual animal(s). Its dimension and configuration must allow for thermal variants, room for exercise, cage furnishings, and establishment of territory if more than one animal is present.

Figure 5*. The outside surface of three sides of these lizard breeding tanks is painted dark green so that the animals cannot threaten each other between tanks.

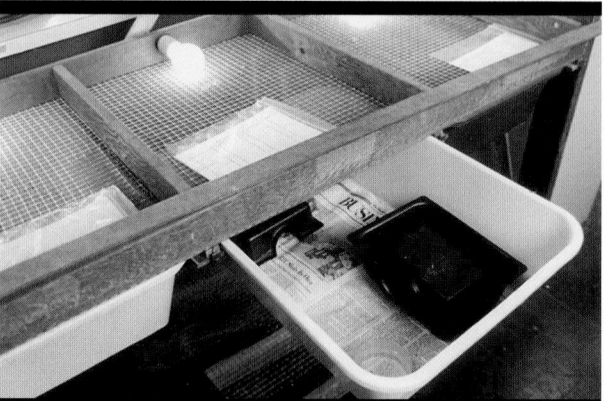

Figure 7*. Two hide boxes, fresh drinking water, and newspaper flooring provide the basic furniture for small reptile enclosures.

Figure 6*. Snakes are secretive creatures. In captivity, they must be able to hide from perceived predators such as cagemates or the human attendant. A secluded place is critical for such a nervous and irritable species, who are reluctant to eat when they feel disturbed. Note the rocks and leaves enabling the snakes to hide.

sand or wood shavings are not suitable for most species because they are often eaten by the animals and may cause digestive impaction. Toads, however, do well with a clean sand substrate deep enough in which to burrow (Gilman, 1984b). Most aquatic amphibians, like newts, frogs, and salamanders, do well with medium-size pebbles and gravel.

Whether using a terrarium or an aquarium to house the animals, seams should be tight and, for reptiles, latches should be placed on all doors and lids. Because the animals will explore or attempt to escape, there should be no rough surfaces on cage walls or seams that can cause abrasions. Excitability and frustrated attempts to escape can also be reduced by eliminating the number of transparent sides on an enclosure. Frogs, for example, when placed into a new tank, tend to jump into glass walls as if the barriers were invisible. By covering the outside of transparent walls with opaque plastic or paint and providing appropriate retreats, the risk of physical trauma can be reduced (Figure 5).

Most amphibians and reptiles are secretive creatures. In captivity, they <u>must</u> be able to hide from perceived predators such as cagemates or the human attendant. A secluded place is critical for nervous and irritable species, such as many snakes, who are reluctant to eat when they feel disturbed (Figure 6). Opaque plastic boxes with small entrance holes make good retreat sites and have the merit of being easily cleaned (Pough, 1991; Figure 7). Hide boxes provide psychological security through containment and familiarity of the immediate environment. They should be available in different zones of the temperature gradient (e.g., warm basking location and cool litter substrate) so that the animal(s) can always find a refuge in a thermocomfortable area of their enclosure. For some species, psychological security is provided by olfactory cues. Snakes and terrestrial salamanders, for example, will make less attempts to escape from their freshly cleaned home cages when a small amount of fecal material is left as a sign of familiarity (Chiszar et al., 1980; Jaeger 1986).

Arboreal and semi-arboreal species should have access to perches on which they can bask and establish a territory, and onto which they can escape. Perch diameter, angle, and placement will depend on the individual animal's size and particular preferences. The availability of perches is a particularly important consideration in the group-housing context. Access to perches will allow the animals to retreat/withdraw to elevated locations, thereby avoiding unnecessary conflicts arising from horizontal space restriction.

- Individual animals may suffer from being maintained in a **social group** if the housing arrangement is not species-appropriate. Investigators must be knowledgeable of the natural, social and territorial tendencies of the species they study to assure that the animals are housed accordingly. When animals are kept with others it is of utmost importance to not only provide adequate space to avoid overcrowding but also to provide sufficient options for basking, feeding and retreating so that there is no reason for competition and no animal is disadvantaged. Solitary species should not be forced to live with conspecifics.

- Frequent, gentle **handling** of non-venomous reptiles can make the individual animal more docile and tractable during minor procedures, such as cleaning of the enclosure, transferring to another area, veterinary inspection and blood collection. Investigators must be informed and take the necessary precautions when handling venomous reptiles.

    Reptiles and amphibians should be handled as quietly as possible with the minimum personnel necessary. Darkened conditions tend to calm the animals and reduce stress reactions. Amphibians should be handled as little as possible because their delicate, moist skin desiccates rapidly thereby making it extremely vulnerable to microlesions.

The above recommendations are intended to meet the minimum conditions necessary to ensure survival and well-being of captive reptiles and amphibians in the laboratory setting. The great diversity of the species will make it necessary for each investigator or instructor to consult the professional and scientific literature (e.g., Bellairs, 1969; Elkan, 1970; Campbell and Busack, 1979; Cowan, 1980; Mattison, 1982; Duellman and Trueb, 1986; Mattison, 1988; Warwick, 1990b; Pough, 1991; Hilken and Willmann, 1994; Warwick and Steedman, 1994; Berry et al., 1995) prior to acquisition of the animals intended for research or teaching purposes, to become well informed on their basic biology, their species-specific environmental and nutritional requirements, and their species-specific behavioral needs. Since there is relatively little published pragmatic information on the successful care of the great variety of amphibian and reptile species, investigators are well-advised to consult curators, zoo keepers and herpetoculturists who have first-hand experience in successful husbandry practices of the species they intend to investigate.

A Note on Salmonella

The Centers for Disease Control has determined that almost all, if not all, reptiles are carriers of *Salmonella* (Mermin et al., 1997). Incidence of salmonellosis related to contact with reptiles has increased in the United States paralleling the increase in households with reptile pets. Because *Salmonella* is only transferred to humans by ingestion, it is strongly recommended that anyone handling reptiles—or surfaces that reptiles may have touched—wash his or her hands immediately after contact. Washing facilities should be in or near the reptile housing facility. Whether or not amphibians are carriers of *Salmonella* is unknown, but the same preventive handwashing is encouraged after handling.

References

Bellairs A 1969. The Life of Reptiles (Volumes 1 & 2). Weidenfeld and Nicolson, London, UK

Berry DJ, Kreger MD, Lyons-Carter JL 1995. Information Resources for Reptiles, Amphibians, Fish, and Cephalopods used in Biomedical Research. National Agricultural Library, U.S. Department of Agriculture, Beltsville, MD

Campbell HW, Busack SD 1979. Laboratory maintenance. In Turtles: Perspectives and Research Harless M, Morlock H (eds), 109-125. Wiley & Sons, New York, NY

Chiszar D, Wellborn S, Wand MA, Scudder KM, Smith HM 1980. Investigatory behavior in snakes, II: Cage cleaning and the induction of defecation in snakes. Animal Learning and Behavior 8, 505-510

Cowan DF 1980. Adaption, maladaption and disease. In Reproductive Biology and Diseases of Captive Reptiles. Society for the Study of Amphibians and Reptiles, Contributions to Herpetology No. 1 Murphy JB, Collins JT (eds), 191-196. Meseraull Printing, Lawrence, KS

Duellman WE, Trueb L 1986. Biology of Amphibians. McGraw-Hill, New York, NY

Elkan E 1970. The management and the pathology of amphibians and reptiles. Veterinary Record 87, 197-199

Gilman J 1984a. Chapter III: Reptiles. In Guide to the Care and Use of Experimental Animals, Volume 2, 19-28. Canadian Council on Animal Care, Ottawa, Canada
Full Text: http://www.ccac.ca/guides/english/V2_84/chiii.htm

Gilman J 1984b. Chapter II: Amphibians. In Guide to the Care and Use of Experimental Animals, Volume 2, 11-17. Canadian Council on Animal Care, Ottawa, Canada
Full Text: http://www.ccac.ca/guides/english/V2_84/chii.htm

Hilken G, Willmann FIF 1994. Preferences of *Xenopus laevis* for different housing conditions. Scandinavian Journal of Laboratory Animal Science 21, 71-80

Jaeger RG 1986. Pheromonal markers as territorial advertisement by terrestrial salamanders. In Chemical Signals in Vertebrates Duvall D, Maller-Schwarze D, Silverstein RM (eds), 191-203. Plenum Press, New York, NY

Jones JP 1978. Photoperiod and reptile reproduction. Herpetological Review 9, 95-100

Mermin J, Hoar B, Angulo FJ 1997. Iguanas and *Salmonella Marina* infection in children; a reflection of the increasing incidence of reptile-associated salmonellosis in the United States. Pediatrics 99, 339-402

Mattison C 1982. The Care of Reptiles and Amphibians in Captivity Blanford Press, Poole, UK

Mattison C 1988. Keeping and Breeding Snakes. Blanford Press, Poole, UK

Pough FH 1991. Recommendations for the care of amphibians and reptiles in academic institutions. Institute of Laboratory Animal Resources (ILAR) News 33(4), S3-S21
Full Text: http://www4.nationalacademies.org/cls/ijhome.nsf

Pough FH 1992. Setting guidelines for the care of reptiles, amphibians and fishes. In The Care and Use of Amphibians, Reptiles, and Fish in Research Schaeffer DO, Klienow KM, Krulisch L (eds), 7-14. Scientists Center for Animal Welfare (SCAW), Bethesda, MD

Warwick C 1987. Effects of captivity on the ethology and psychology of reptiles. Herpetoculturist 1, 10-12

Warwick C 1990a. Important ethological and other considerations of the study and maintenance of reptiles in captivity. Applied Animal Behaviour Science 27, 363-366

Warwick C 1990b. Reptilian ethology in captivity: Observations of some problems and an evaluation of their aetiology. Applied Animal Behaviour Science 26, 1-13

Warwick C, Steedman C 1994. Naturalistic versus clinical environments in husbandry and research. In The Health and Welfare of Captive Reptiles Warwick C, Frye FL, Murphy TB (eds), 113-130. Chapman & Hall, London, UK

Wright K 1994. Acclimation-Maladaptive Syndrome in captive amphibians. The Vivarium 6(3), 12-13

The views expressed here are those of the author and do not represent views of the U.S. Fish and Wildlife Service.
Michael Kreger is a biologist with the U.S. Fish and Wildlife Service (USFWS) Division of Scientific Authority. Prior to joining USFWS, he was a technical information specialist at the Animal Welfare Information Center of the U.S. Department of Agriculture where he specialized in non-mammalian research animals and zoo/aquarium issues. His Masters degree involved studying the physiology and behavior of phytons and skinks in response to handling. He is currently completing a Ph.D. focusing on behavior and survival of reintroduced whooping cranes.

*Photo has been taken by the author at the National Zoological Park's Reptile House in Washington, DC.